The central role of soil chemistry in the ecosystem is b
evident. The effects of increased levels of atmospheric
the accelerated use of pesticides, on soil fertility, for e:
focus of much high level debate.

This text begins by defining the relationship between soil chemistry and
other fields as diverse as plant science and pollution science. A detailed
description of the components of soil follows, including inorganic, mineral
and organic matter. The book addresses cogent issues such as soil fertility
and soil pollution. In a concluding chapter, a review of future analytic
advances in the study of soil chemistry is given, emphasising the importance
of the soil chemist in equitable and sustainable land use and agricultural
policy.

The book is an ideal starting point for the student undertaking
undergraduate study in the environmental and soil sciences.

SOIL CHEMISTRY AND ITS APPLICATIONS

CAMBRIDGE ENVIRONMENTAL CHEMISTRY SERIES
Series Editors
Professor P. G. C. Campbell, *Institut National de la Recherche Scientifique, Université du Québec*
J. N. Galloway, *Department of Environmental Sciences, University of Virginia, USA*
R. M. Harrison, *Department of Environmental Health, School of Biological Sciences, Birmingham University*

Other books in this series:
1 P. Brimblecombe   *Air Composition & Chemistry*
2 M. Cresser and A. Edwards   *Acidification of Freshwaters*
3 A. C. Chamberlain   *Radioactive Aerosols*
4 R. M. Harrison, S. J. de Mora, S. Rapsomanikis and W. R. Johnston
   Introductory Chemistry for the Environmental Sciences

SOIL CHEMISTRY AND ITS APPLICATIONS

○ ○ ○ ○ ○ ○ ○ ○ ○ ○ ○ ○ ○ ○ ○ ○ ○

MALCOLM CRESSER &
KEN KILLHAM
Department of Plant & Soil Science
University of Aberdeen

TONY EDWARDS
Macaulay Land Use Research Institute
Aberdeen

Published by the Press Syndicate of the University of Cambridge
The Pitt Building, Trumpington Street, Cambridge CB2 1RP
40 West 20th Street, New York, NY 10011-4211 USA
10 Stamford Road, Oakleigh, Melbourne 3166, Australia

© Cambridge University Press 1993

First published 1993
Reprinted 1995

A catalogue record for this book is available from the British Library

Library of Congress cataloguing in publication data

Cresser, Malcolm S.
Soil chemistry and its applications / Malcolm Cresser & Ken Killham, Tony Edwards.
   p. cm. – (Cambridge environmental chemistry series)
Includes bibliographical references and index.
ISBN 0-521-32269-3. – ISBN 0-521-31134-9 (pbk.)
1. Soil chemistry.   2. Soil fertility.   3. Soil pollution.
I. Killham, Ken.   II. Edwards, Anthony.   III. Title.   IV. Series.
S592.5.C74   1993
631.4'1–dc20   92-30511   CIP

ISBN 0 521 32269 3 hardback
ISBN 0 521 31134 9 paperback

Transferred to digital printing 1999

CONTENTS

	Preface	xiii
1	**Why study soil chemistry?**	1
	Soil chemistry and plants	2
	Soil chemistry and soil biology and biochemistry	3
	Soil chemistry and soil physics	3
	Soil chemistry, geochemistry and soil formation	3
	Soil chemistry and water chemistry	4
	Soil chemistry and pollution science	4
	Soil chemistry as a subdivision of chemistry	5
	Soil chemistry in history and pre-history	5
	The nature of soil – what it is and what it does	7
	The transition from rock to soil	8
	The remit of the soil chemist	8
2	**Inorganic and mineral components of soils**	10
	The nature of rocks	10
	Igneous rocks	11
	Primary and secondary minerals	14
	Sedimentary rocks	14
	Diagenesis	16
	Geochemical weathering	17
	Silicates	18
	Simple tetrahedra	18
	Chain silicates	19
	Sheet silicates	20
	Framework silicates	21
	Other important sheet structures	22
	Isomorphous substitution	23
	2:1 and 1:1 type minerals	23
	Important minerals other than silicates	24
	Clay minerals	24
	Interstratified minerals	26
	Allophane	26
	Imogolite	27
	Soil minerals and soil chemical reactions	27

Contents

3 Soil organic matter — 28
Introduction — 28
Components of soil organic matter — 28
   Decomposing residues of plants — 29
   Cellulose — 31
   Hemicelluloses — 32
   Lignin — 34
Soil biota and organic matter turnover — 35
   The soil decomposer community and microbial succession — 36
   Kinetics of organic matter breakdown — 37
Resistant soil organic matter — 38
   Chemically protected organic matter — 39
   Physically protected organic matter — 41
Detailed characterisation of soil organic matter — 42
   Chemical fractions of soil humus — 43
   Humic acid — 43
   Fulvic acid — 44
   Humin — 44
Soil organic matter and soil structure — 44
   Organic matter as a transient binding agent — 45
   Organic matter as a temporary binding agent — 45
   Organic matter as a persistent binding agent — 46
   Organic matter and soil structural stability — 46
Rates of degradation and accumulation of organic matter in soil — 47
   Turnover of soil organic matter — 47
   Accumulation and loss of soil organic matter — 48
The organic chemistry of anaerobic soils — 49
   Soil organic matter and the generation of anaerobic conditions — 49
   Anaerobic microsites in soil — 52
   Competition between electron acceptors in anaerobic soils — 53
   Consequences of anaerobiosis in soils — 53
Soil organic matter and trace element availability — 54
   Release of trace elements into soil solution — 54
   Depletion of trace elements in soil solution by bonding to organic matter — 55
   Chelation by organic ligands — 56
Other roles of soil organic matter — 57

4 Soil chemical reactions — 58
Cation-exchange properties of soil clays — 58
   Inner-sphere and outer-sphere surface complexes — 59
   Measurement of exchangeable cations and exchange capacity — 59
Some theoretical aspects of cation exchange — 61
   The Ratio Law — 61
Particle size distribution in soils — 62
Soil pH — 65
   Measurement of soil pH — 66
   The transient nature of soil pH — 67
Soil pH and nutrient availability — 68
   The base cations — 69
   Copper, zinc and cobalt — 70

Iron and manganese	72
Molybdenum	73
Phosphorus	74
Boron	77
Nitrogen	77
Sulphur	78
Other essential elements	78
Saline and sodic soils	79
Saline soils	79
Sodic soils	82
Saline-sodic soils	83
Gypsum in soil	83
Oxidation–reduction reactions in soils	83
The relationship between Eh *and* (pe + *pH*)	88
The solubility of iron and manganese under reducing conditions	88
Reducing conditions and soil pH	90
Chemical problems associated with anaerobic conditions	90
Biogeochemical cycling of nutrient elements	90
Cycles with no gaseous components	90
Cycles with gaseous components	91
Hydrogen	91
Oxygen	92
Carbon	92
Nitrogen	93
Sulphur	93
Chlorine	95
Selenium	95
Methods for studying element cycles	95
Precipitation inputs	97
Dry deposition inputs	97
Litter	97
Litter decomposition	98
Outputs in drainage waters	98
Gaseous fluxes	98
Animal wastes	98
Microbial cycling	98
Rain simulation techniques	99
General comments	99
5 Soil fertility	**100**
What is soil fertility?	100
What influences natural fertility?	101
Management options and soil fertility	103
Crop rotation	103
Intercropping	105
Unavailable nutrient pools	105
Plant breeding and selection	106
Use of fertilisers	106
Nitrogen fertilisation	108
Phosphorus fertilisation	111
Potassium fertilisation	112

Contents

Soil acidity	113
Soil pH and its measurement	114
Liming material	116
Acidity and plant growth	116
Trends in soil pH	119
Effects of liming on soil solution chemistry	121
Spatial pH variations	122
Evaluation of soil fertility	123
Soil analysis	124

6 Soil chemistry and freshwater quality — 128

The hydrological cycle	130
Factors affecting hydrological pathways	130
Rainstorms, snowmelt and freshwater quality	131
Buffering of pH in upland streams	134
Catchment characteristics influencing soil–water interactions	136
Precipitation characteristics	136
Other climatic factors	138
Soil characteristics	140
Soil parent material	141
Catchment characteristics	142
Modelling water quality	143
Vegetation effects	147
Improved drainage and water quality	147
Agrochemicals, soils, and freshwater quality	148
Nitrogenous fertilisers and manures	149
Other fertilisers	149
Liming materials	150
Pesticides	150

7 Soils and pollution — 151

Sources of soil pollution	152
Sludges and animal wastes	152
Fertilisers	153
Atmospheric inputs	155
Radionuclide deposition	155
Ammonia	157
Soils and acid deposition	157
Effects of increasing carbon dioxide	162
Adsorption of pesticides by soils	163
Types of pesticide adsorption to soil organic matter	163
Van der Waal's forces	163
Hydrogen bonding	164
Hydrophobic bonding	164
Ion exchange	164
Charge transfer	164
Ligand exchange	164
Chemisorption	165
Organic matter fractions involved in pesticide adsorption	166

8	**The future of soil chemistry**	168
	The changing nature of research in soil chemistry	168
	Soil as a component of the whole ecosystem	169
	New research tools in soil chemistry	171
	Soil chemistry and policy makers	172
	References	173
	Index	186

PREFACE

Over recent years, there has been a remarkable rise in the perceived importance of soil chemistry. A couple of decades ago, it tended to be regarded either as a small subdivision of agriculture and horticulture, or as a subject studied for its own interest by a handful of perhaps slightly eccentric physical or organic chemists. Today it is regarded by many as a subject at the core of environmental science. The change has been brought about largely by the worldwide recognition of the vital (and much wider) role played by soil, in controlling the air we breathe, the water many of us drink, and the extent of damage by pollution. In soil chemistry textbooks, however, tradition has largely prevailed, and the main thrust has been on discussion of soil as a medium for plant growth.

The authors decided that the time had come for a rather different approach, and set about preparing this volume. Their aim was to introduce students to the nature of soils, soil processes and soil formation, but in a much broader context than soil fertility alone.

Of necessity they have had to be selective, and no doubt in so doing will not satisfy all potential users of the book. However they hope that the message will get across to students that an understanding of soil chemical behaviour is crucial in the modern world, and that soil is dynamic, so that soil and soil problems are constantly changing. Over coming decades, soil chemistry will grow even more in importance.

The authors are indebted to many students and colleagues who, by their enthusiasm, questions or doubts over many years have lead to the evolution of this text, and to Peter Campbell for many helpful suggestions. They would welcome feedback from users at any level for improvements to a possible second edition.

Aberdeen, 1992
Malcolm Cresser,
Ken Killham,
Tony Edwards.

1

Why study soil chemistry?

It doesn't matter who you are or where you are, your very existence and survival depend upon the chemical reactions taking place all the time in soil. Indeed, this book itself owes its existence to the same reactions. The paper originates from trees grown in soil, the ink is from chemicals traceable back to the soil, and the authors are mere links in a complex food chain passing through soil-based cycles. The longevity of their role depends upon soil-derived food and regulation of the composition of the atmosphere by chemical reactions controlling the growth of photosynthesising plants. Of all the scientific disciplines that interact to make up the complex web of environmental science, soil chemistry could therefore be regarded as the most central. It is worth elaborating upon this concept, because any vague and fuzzy images the reader might have of possible applications of soil chemistry should then spring sharply into focus.

Figure 1.1 is an attempt to demonstrate the central role of soil chemistry, and its major interactions with other aspects of science. The outer ring of topics is impressive in its scope, but is by no means exhaustive. We could, for example, include terrestrial and aquatic zoology, since soils affect plants and water quality, and hence animal life. Plants may influence climate on a local scale and, in the long term, on a global scale, so we could add meteorology. Soil chemistry regulates soil fertility, so its links to agriculture and horticulture are obvious. Less obvious perhaps are its links with archaeology, social and economic history, and human geography, which arise as a consequence of the need for sustainable agriculture. We will only consider briefly here the major disciplines included in Fig. 1.1, and leave the reader to ponder over omissions, whether accidental or by design.

2 Why study soil chemistry?

Soil chemistry and plants

Soil chemical properties regulate the availability of essential major elements (nitrogen, phosphorus, sulphur, potassium, calcium and magnesium) and trace elements (boron, copper, iron, manganese, molybdenum and zinc). 'Essential' in this context describes an element which must be present if the plant is to proceed through every physiological stage of its growth cycle. Soil chemical characteristics also govern the availability of elements such as cobalt which may be essential from the viewpoint of animal dietary requirements, and of elements or ionic species such as selenite which may be toxic to higher plants or animals. The regulatory mechanisms involved are discussed in Chapter 4. Soils are also important in restricting undesirable side effects from plant protection chemicals such as selective herbicides, fungicides, molluscicides, etc. (see Chapter 7).

Fig. 1.1 Schematic representation of the interactions between soil chemistry and other branches of soil science and environmental science.

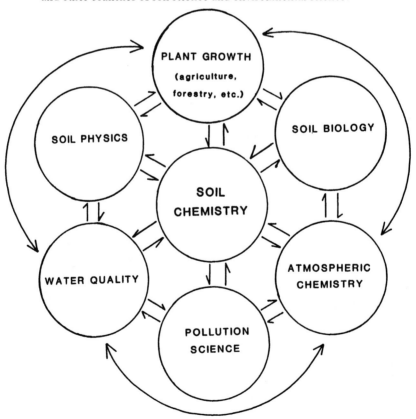

Plants, in turn, may substantially influence soil physicochemical properties in many ways. The major mechanisms include soil acidification by release of hydrogen ions at the root to compensate for plant uptake of base cations (calcium, potassium, magnesium and sodium), anion uptake, modification of soil moisture content, the effects of organic compound exudation from roots, root respiration, incorporation of plant litter and its subsequent degradation in the soil, and the effects of organic compounds leached from foliage or litter. These and other influences are discussed where appropriate in later chapters.

Soil chemistry and soil biology and biochemistry

As will be seen in Chapters 2–5, the biological population of the soil, especially the microbial population, plays a vital role in the biogeochemical cycling of nutrient elements such as carbon, nitrogen, phosphorus and sulphur, and hence in the regulation of soil fertility. However, the activities of macro- and micro-organisms in soils are themselves regulated by soil chemical properties. The plant species growing in a soil may influence soil biological activity both indirectly, through effects upon soil chemistry, and directly, through the size and biodegradability of the litter load. Soil physical properties, especially drainage status, may also considerably influence biological activity.

Soil chemistry and soil physics

The chemical reactions which occur in soil may have far-reaching effects upon the physical properties of the soil, such as the stability of its structural units, especially in regions with arid climates. In many soils in wetter areas, the particle size distribution, which governs its drainage characteristics, is a function of the chemical weathering of the minerals present in the soil parent material. Soil organic matter also plays an important role in water retention, and in the establishment of stable soil structure, the macropores created facilitating drainage. As will be seen in Chapter 4, soil drainage is crucial if the soil is to remain well aerated and plant growth problems associated with anaerobiosis are to be avoided.

Soil chemistry, geochemistry and soil formation

The chemical nature of the minerals initially present in the rock, till or sediment from which a soil has developed exerts a substantial influence upon the soil chemical properties at any given time in its evolution. These properties may, in turn, have a striking effect upon the soil biological activity, and the combined biological and chemical effects govern the type of soil which is formed and the uses to which it may be put. We will return briefly to the subject of soil formation later in this chapter.

Soil chemistry and water chemistry

Except where outcropping rock dominates the landscape, precipitation (rain, sleet, snow, etc.) reaching the soil surface flows through or over soil before reaching groundwater or draining into streams or lakes. During this period of soil–water contact, a range of chemical reactions take place which regulate the chemical composition of the freshwater eventually obtained. Provision of freshwater supplies of adequate quality for an ever-increasing world population is such an important topic that Chapter 6 has been devoted in its entirety to this theme. Solute and suspended solids are transferred by rivers to the oceans, so soil chemical reactions also play a role in marine chemistry.

Precipitation itself contains a significant amount of solute which will also interact with the soil solid components as the water drains through or over soil. The solute may be of natural origins, for example, sea salts from oceanic spray, sulphuric acid from volcanic sulphur dioxide, etc., or it may be a consequence of human activities, i.e. pollution. The extent to which soil modifies the passage of these inputs to waters depends upon the stage of soil evolution. This may be true because the hydrological pathway followed by the water may change with the degree of soil development, or simply because of the changes which occur over decades and centuries in soil physicochemical characteristics.

Soil chemistry and pollution science

The world's soils serve as a depository for vast quantities of pollutants. Sometimes the fate and consequences of individual pollutants become very emotive topics, for example nuclear waste or fallout, acid rain or lead from petrol. Another important source of potential pollution is the spreading of manures and sewage sludge on soil, either for disposal or for their fertiliser effect. Such materials are a major component of the annual phosphorus budget in the UK, for example. The ultimate fate of these pollutants, and of fertilizers and other agrochemicals added to soils, is an increasingly important area of study by soil chemists. Of particular concern is the possibility of toxic substances entering the human food chain in unacceptable amounts. Many soils have an extraordinary capacity for rendering pollutants innocuous, but at the present time there is much concern that this capacity is being abused to the extent that critical loads may be exceeded. 'Critical' is the term applied to the annual load of a pollutant at which the normal functioning of the ecosystem breaks down in some way.

Atmospheric pollutants may influence plant growth directly, and thus exert an indirect effect upon soil chemistry, for example by changing the plant litter load or degradability, or the soil moisture content. Forest growth may also influence the efficiency with which atmospheric pollutants are captured.

Soil chemistry as a subdivision of chemistry

From the brief summary above of the role of soil chemistry in the broader context of environmental science, it should be immediately apparent that the subject draws upon knowledge of physical, organic, inorganic and analytical chemistry. Developments in analytical science over the past two to three decades have made a very significant contribution to our understanding of the complex workings of the soil–plant–water ecosystem. However the soil chemist must avoid thinking of soil chemistry in isolation, and requires at least a rudimentary knowledge of the interacting areas included in Fig. 1.1. Thus, although this book is essentially a chemistry text, the authors have attempted throughout to put soil chemistry in a broader sphere of reference. By avoiding making the text unnecessarily complicated, they have also tried to produce a book which will serve as an interesting and readable introduction to this fascinating topic for other types of environmental scientist.

Soil chemistry in history and pre-history

As an experimental science, soil chemistry is quite ancient insofar as manures and primitive liming materials such as shell sands have been employed to improve soil fertility for thousands of years. Although detailed investigation of archaeological agricultural sites is perhaps less glamorous than excavation of dwellings, certainly there is evidence to suggest that the contents of domestic waste pits were distributed over wide areas. To a degree, the interpretation of a wide spread of pottery, tool and shell fragments over field sites must remain speculative, although alternative interpretations of such labour-intensive practices are not easy to come by. Certainly it is not difficult to conceive that, following the move from hunting and gathering through slash and burn to more static agricultural areas with identifiable field boundaries, soil infertility problems would soon have arisen, even with primitive, low-yielding cereal varieties. It is a small jump from there to the point at which some early farmer would notice improved growth in the vicinity of animal waste, and perhaps try a few experiments.

Certainly by the birth of Christ, it is possible to find mention of the use of manuring as if it was a long-standing practice (see e.g. Varro's *Res Rusticae*, ca 37 BC or Columella's *Res Rustica*, ca 60 AD). Purists might argue that this muck-and-magic agriculture should not be described as chemistry. However, it is relevant because tried and tested agricultural practices should have pointed the first true soil chemists in the right direction in the seventeenth century. In fact, the exact opposite happened. Once it was established that a potted plant could increase in weight with nothing apparently added but water, it was erroneously concluded that water was all that the plant required, plant organic components coming from soil organic matter. Thus the manure provided the organic matter in the plant. This misconception is hardly surprising, for it would have required a truly exceptional mind to conceive the idea of fixation of a gaseous component of the air at that stage. It survived for several decades, however, with minor modifications as it was demonstrated that substances such as saltpetre could sometimes dramatically accelerate crop growth. An excellent and concise account of this early work has been presented by Russell (1987) in his celebrated text *Soil Conditions and Plant Growth*. It was not until 1782 that Senebier concluded that fixation of an atmospheric component in the presence of light (i.e. photosynthesis) was the origin of plant organic matter, a hypothesis eventually proved conclusively by de Saussure (1804). De Saussure also demonstrated the importance of soil for supplying nitrogen, potassium and phosphorus to plants, thus sowing the seeds for modern soil chemistry.

Widespread belief in soil humus as the source of plant carbon remained until Liebig in 1840 published a highly critical attack upon the prevalent attitudes and concepts. He stressed the importance not just of photosynthesis, but also of a range of plant nutrients, emphasising that the lack of any one essential nutrient could be sufficient to restrict plant growth. The next major advance came some 37 years later with developments in bacteriology. Schloesing and Muntz (1877) were able to demonstrate that ammonium in sewage could be converted to nitrate, after an initial delay, by bacteria, and it was soon shown that the same process occurred in soils. Since that major advance, progress has been slow but steady, with more and more detail being filled in about both the biological and the chemical reactions occurring in soils and the interactions of soils with other ecosystem components. The current state of our knowledge of soil chemistry arising from this progress constitutes the basis of Chapters 2 to 7.

The nature of soil – what it is and what it does

Soil may be defined as material of variable depth with a substantial solids content at the Earth's surface which is undergoing change as a consequence of chemical, physical and biological processes. Thus, towards the bottom of a deep soil pit, material that was not changing with time would be classified as parent material, but not as soil.

Soil essentially consists of three phases, a solid phase, a solution phase and a gas phase. The solid phase usually includes an intimate mixture of mineral material, originating from rock, sediment or till, and organic material arising as a consequence of biological activity. It interacts continuously with the solution phase, which originates from precipitation infiltrating the soil or from rising water or water moving laterally. The chemical composition of the soil solution depends upon the physicochemical characteristics of the soil solids, precipitation solute composition, biological activity within the soil matrix, and to some extent upon contact time. It contains both organic and inorganic components. The gas phase, or soil atmosphere, composition depends upon biological activity also. It may be greatly enriched in carbon dioxide (up to *ca* 3–4%) compared to normal above-ground air (*ca* 0.035%) as a consequence of microbial and root respiration, and relatively depleted in oxygen. Under certain conditions, it may contain significant amounts of gases such as nitrous oxide or ammonia, and even hydrogen sulphide and ethylene. These are discussed fully in Chapter 4.

That the soil fulfils a multiple role in most ecosystems should already be clear from the discussion of the relationship between soil chemistry and other branches of environmental science. It is worth summarising briefly here the major roles, however. Soil provides plants with essential major and minor nutrients, with water and with firm anchorage. It acts as a sink for organic detritus (plant and animal remains and waste products) and for natural and pollution inputs from the atmosphere.

Unless subjected to excessive pollution stress, it regulates the solute chemistry of freshwaters to make the appropriate environments to support aquatic life. If properly managed, soil continues to fulfill all these roles. Proper management requires an understanding of soil chemistry.

The transition from rock to soil

We are now in a position to consider briefly the processes involved in the transformation from rock to soil. Clearly rock will be subjected to physical weathering as a consequence of the action of wind, rain, frost and rapid temperature changes. Some dissolution may also occur, releasing essential nutrient elements such as calcium, potassium, magnesium and phosphorus and the various trace nutrient elements. However nitrogen, a major nutrient element, is not released in significant amounts by mineral weathering, although small inputs of ammonium- and nitrate-nitrogen occur even in pollution-free rain. The rock surface tends to be colonised therefore by lichens. Lichens consist of algae and fungi living in a mutually beneficial (symbiotic) relationship. The algae can fix atmospheric nitrogen and photosynthesise. They are thus able to provide the organic substrate for the growth of the fungi. The fungi, in turn, speed up the chemical attack of the rock surface by organic acid production, thus providing more nutrients for the algae. In this way a thin soil layer starts to accumulate. The organic matter content starts to build up, increasing the moisture holding capacity of the soil to the point where other species such as mosses may become established. Initially the skeletal soil may be subject to erosion, except in cracks in the rock. Over a long timescale, however, the soil will spread, especially if the soil becomes stabilized by the establishment of vegetation cover. Chemical reactions will penetrate deeper into the rock, as the evolving soil is mixed and remixed by soil biota.

Subsequent soil evolution depends upon a number of factors, the most important being time, precipitation amount and temporal distribution, temperature, topography and aspect. We shall return to the topic of soil formation under diverse climatic conditions where appropriate in Chapters 2–4.

The remit of the soil chemist

Clearly the remit of the soil chemist is very broad. She/he may work in agriculture, forestry or horticulture, or for a water quality control authority. Interest may be centred around optimisation of plant growth, provision of adequate grazing, pollutant effects upon crop yields or freshwater quality, or the passage of radioactive fallout into the food chain. These examples should suffice to give some insight into the scope of the subject.

2

Inorganic and mineral components of soils

As we saw in Chapter 1, rock plays a vital role in soil formation, whether the soil evolves from solid, outcropping rock weathering *in situ* or from fragments which have broken away from massive outcrops and been transported, often down slope. This movement occurs as a consequence of the action of forces arising from gravity, wind, rain, ice movement or drainage water, either alone or in combination. It may result in a substantial degree of mixing of parent materials. Clearly the origins, and hence the chemical composition, of the parent rock might be expected to influence significantly the chemical and physical properties of the evolving soil at any stage in its development, and this is indeed the case. In the present context we need to consider the properties of rocks insofar as they influence the characteristics of the soils ultimately obtained, and the transformations which rock and rock-derived products undergo over the timescale from seconds to tens of centuries resulting from interactions with water and the soil flora and fauna.

The nature of rocks

It is intuitively obvious that rocks are chemically very stable. If they were not, clearly chemical attack upon rock outcrops would be much more rapid than it is. It follows that the chemical bonding in the macromolecules (minerals) that constitute rocks must produce lattices which are very favourable in energetic terms. Disruption of the lattice, i.e. dissolution, requires a high input of energy to overcome the forces holding the atoms together into well-defined mineral lattices.

The core of our planet exists at very high temperature and pressure. Although never sampled directly, it is thought to have a composition not unlike that of iron meteorites and a mean density of around 11 000 kg m^{-3}, so that it is 11 times more dense than water (Bowen,

1979). The core is surrounded by the mantle, which has a mean relative density of 4.5, and extends to about 2900 km below the surface. It is believed that material from the upper mantle occasionally reaches the surface from volcanic activity in the vicinity of the mid-Atlantic and mid-Pacific ridges, where the Earth's outer crust is thinnest. However the rocks that are eventually transformed into soils generally originate in the outer crust. The crust is very heterogenous, but has a mean density of 2800 kg m^{-3} (relative density, 2.8). The rocks that dominate the crust fall into three broad groups, igneous rocks (those which have crystallised directly from molten magma), sedimentary rocks (those which have formed from deposited sediments), and metamorphic rocks. This latter group is formed from igneous or sedimentary rocks which have been subjected to elevated temperatures and pressures. Under such conditions they have undergone partial recrystallisation and solid state reactions to yield more dense, metamorphic rocks.

Igneous rocks

The crystallisation processes which occur as molten magma slowly cools tend to follow a reasonably well-defined path (Bowen, 1979; Mason, 1982). The first product is olivine ($MgSiO_4$), a dense mineral which settles out to form a basalt. This is followed by minor amounts of iron-rich minerals such as pyrite (FeS_2), pyrrhotite (FeS), magnetite (Fe_3O_4), chromite ($FeCr_2O_4$) and ilmenite ($FeTiO_3$). These may be accompanied by insoluble sulphides of other transition metals such as cobalt, copper, nickel and zinc. The presence of some of these transition metals leads to very intensely coloured mineral particles, so that the resulting rocks are often quite dark in colour.

As solidification proceeds, a continuous series of rocks evolves with progressively less and less aluminium and, particularly, magnesium, and a steadily increasing content of silicon. Thus the first group formed, known collectively as the ultrabasic rocks because of their relatively high content of base cations, contains <45% of silicon dioxide. It includes fine-grain basic olivine basalts, medium-grain basic dolerites, and coarser grain peridotite and serpentinite (Landon, 1991). In addition to olivine, these rocks often contain calcic plagioclase (on the calcium side of the feldspar series $Na_2O.Al_2O_3.6SiO_2 - CaO.Al_2O_3.2SiO_2$) and lesser amounts of enstatite ($MgO.SiO_2$), augite ($CaO.2(Mg,FeO).(Al,Fe)_2O_3.3SiO_2$), biotite ($K_2Al_2Si_6(Fe^{2+},Mg)_6O_{20}(OH)_4$), and magnetite (Fitzpatrick, 1980; Landon, 1991).

Rocks with 45–55% silicon dioxide are known as basic rocks. They include fine-grain basalt, medium-grain dolerite and medium to coarse-grain gabbro. All are abundant in plagioclase, and may contain variable amounts of olivine, ilmenite, augite, and magnetite, with small amounts of other minerals (Fitzpatrick, 1980).

Intermediate igneous rocks contain 55–65% silicon dioxide. This series includes fine-grain andesite, medium-grain plagioclase porphyries and diorites, amphibolites and syenites. The rocks contain predominantly plagioclase feldspars, with variable, and often small, amounts of a variety of minerals such as biotite, quartz (SiO_2), augite, and hornblende ($4Ca_3Na_2(Mg,Fe)_8(Al,Fe)_4Si_4O_{44}(OH)_4$).

Acid igneous rocks contain >65% silicon dioxide, and therefore invariably contain free quartz. They include: obsidian, generally a black, glassy rock showing conchoidal fracture; fine-grain rhyolite, a glassy or microcrystalline mass of alkali feldspar ($(Na,K)_2O.Al_2O_3.6SiO_2$) and quartz, often with small crystals of orthoclase and other minerals; granites, coarse-grain rocks dominated by alkali feldspar and quartz, with small amounts of plagioclase, biotite and muscovite ($K_2Al_2Si_6Al_4O_{20}(OH)_4$), and traces of other minerals; and microgranites, similar in composition to granites, but finer-grained.

To the reader with little or no knowledge of geology or mineralogy, the above overview of the constitution of igneous rocks, summarised even more concisely in Fig. 2.1, may appear to be disturbingly complex, especially when it is pointed out that the lists are (intentionally) not exhaustive. It is worthwhile at this point therefore to give some indication of the value to the soil chemist of a knowledge of rock types. When biogeochemical cycling and soil formation are discussed later in this chapter and in Chapter 4, it will become clear that, in the long term, the acidity and fertility of all natural (uncultivated) soils depend upon the rate of release of nutrient cations from soil minerals and the other factors of soil formation mentioned already in Chapter 1. An ability to identify bedrock or rock fragments (stones) in soil is therefore a useful guide to potential long-term fertility. This is especially true when soils with different parent materials at sites with similar climate, age, and topography are being compared. Such a comparison has been carried out in the authors' laboratories for two catchments in northeast Scotland, for example (Edwards *et al.*, 1985). Soil acidification is much more advanced for soils derived from granite than it is for soils derived from a more basic rock, a quartz–biotite–norite. From analysis of the parent materials and measurement of the geochemical weathering rates, it is

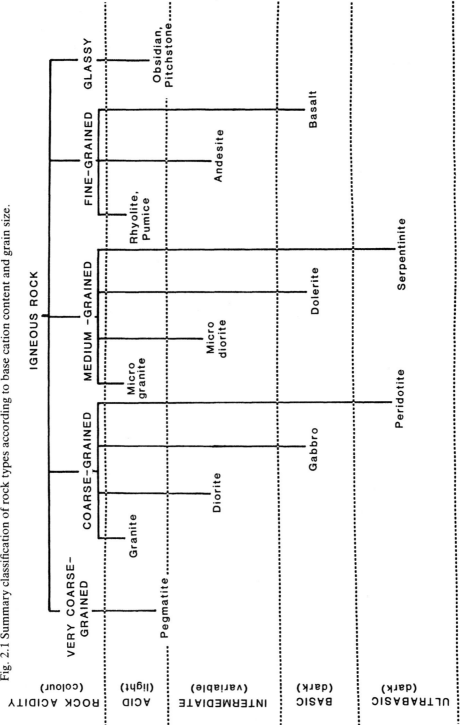

Fig. 2.1 Summary classification of rock types according to base cation content and grain size.

possible to predict how long it will be before the soils in the more base-rich catchment are susceptible to serious acidification. Identification of rock fragments may also be important in establishing the parent material and age of soils.

Primary and secondary minerals

From the previous section it should be immediately apparent that a large number of different minerals may be encountered in soils evolved from igneous rocks. In practice 30 or more is not atypical, although the number of very abundant minerals is usually much less. The most commonly found minerals and their physical properties are listed in Table 2.1. So far, however, we have only considered the primary minerals, that is the minerals which were contained in the parent rock. It will be seen later that soils also contain minerals which are formed as a consequence of chemical reactions occurring within the soil profile. These may be precipitated from solute in the soil solution, sometimes as coatings around existing mineral particles to yield a very high surface area of the mineral forming, or they may be solid degradation products from the partial decomposition of primary minerals. They are known as secondary minerals. Their formation and importance are discussed later in this chapter.

Sedimentary rocks

Sedimentary deposits can take many forms. Landon (1991), for example, lists 14 sediment types of relevance to soil chemistry. A similar approach has been adopted here in Fig. 2.2, which shows sedimentary formations of chemical and biological origins, and Fig. 2.3, which lists sediments resulting from gravitational settling in water courses, lakes or oceans. Further subdivision according to cementing agents is also possible (Landon, 1991), but is unnecessary here. In this text, for simplicity, it is convenient to consider only five major groups:

(1) transported unchanged minerals (in water or by wind);
(2) secondary minerals formed in water;
(3) precipitated calcium carbonate or silica of marine organism origin;
(4) deposits (mainly gypsum or salt) resulting from evaporation in arid climates;
(5) organic matter deposits.

Table 2.1. *Characteristics of minerals commonly found in igneous rocks (after Fitzpatrick (1980) and Landon (1991))*

Mineral	Physical characteristics	Typical composition, %					
		SiO$_2$	Al$_2$O$_3$	FeO	MgO	CaO	K$_2$O
Quartz	Hard, white and colourless. Often occupies interstices between other minerals in igneous rocks. Conchoidal fracture.	100					
Orthoclase	White to pink crystals, often large and equidimensional.	65	20	<1		1	8
Plagioclase	White, often lath-shaped crystals.	53	30			12[a]	<1
Muscovite	Platy colourless crystals, often with pearly lustre on cleavage surfaces. Soft, with perfect basal cleavage.	45	37	<1	<1		10
Biotite	Similar to muscovite, but brown, with dark, metallic lustre.	36	18	22	7	<1	9
Olivine	Yellow, olive green or brown, hard crystals with conchoidal fracture.	35	<1	61	3	1	
Augite	Dark green to black crystals.	47	3	20	7	17	<1
Hornblende	Very similar to augite.	45	11	13	10	12	1

[a] Composition cited is for a calcic plagioclase, labradorite.

Inorganic and mineral components of soils

Diagenesis

Diagenesis is the name given to the complex series of interacting physicochemical processes whereby sediments are converted to sedimentary rocks. The changes are characterised by a reduction in moisture and organic matter contents and physical hardening with the increase in density (Bowen, 1979). Transported mineral deposits, for example, may be cemented by silica. Secondary clay minerals may be converted to shales by partial dehydration and chemical changes occurring under pressure. Calcium carbonate deposits may be converted to limestone, or to dolomite (calcium magnesium carbonate) by exchange of magnesium ions for calcium ions.

The transported minerals deposited in lakes and oceans are often those that are highly resistant to dissolution, such as quartz, corundum (Al_2O_3) and hematite (Fe_2O_3). If these mineral crystals are then cemented together by quartz, the resultant sandstone rock is extremely acidic, being very high in quartz and low in base cations. The soils evolved from such sandstones and shales may be expected to be susceptible to soil acidification in the long term (see Chapter 7).

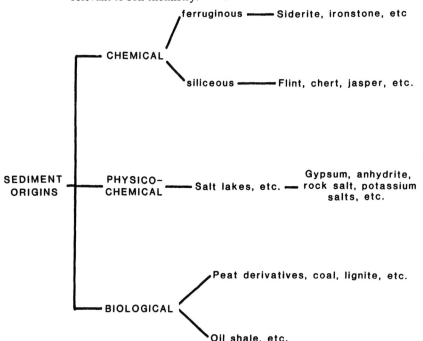

Fig. 2.2 Summary of main sediment types of chemical and biological origin relevant to soil chemistry.

Geochemical weathering

As we saw in Chapter 1, in the early stages of soil formation most of the nutrient elements essential for plant growth come directly from weathering minerals in rocks. In Chapters 4 and 5, the importance of geochemical weathering in the maintenance of soil fertility will also become apparent. From the preceding discussion of the nature of rocks, it should be clear that rocks of diverse origins will contain different reserves of nutrient elements, especially of the base cations calcium, magnesium, potassium and sodium. An understanding of geochemical weathering rates of different minerals is, however, just as important as a knowledge of the chemical composition of the minerals, which may in any case be variable for some elements from site to site. Data on geochemical weathering rates provide information about the rates at which weathering soil minerals may replenish nutrient elements removed from a soil in drainage waters or harvested crops.

The clearest insight into the very wide range of mineral solubilities

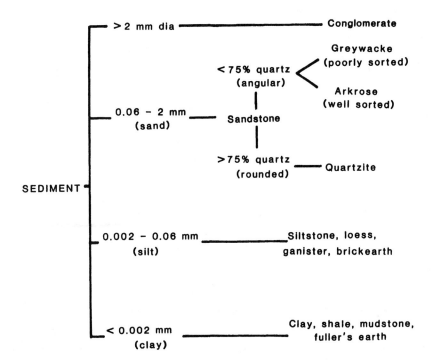

Fig. 2.3 Summary of main sediments resulting from gravitational settling in water.

comes from examination of the variability of the solute composition of riverwaters from drainage basins containing soils derived from different rock types. Walling and Webb (1981), for example, list annual outputs of calcium, magnesium, sodium and potassium from 47 British catchments. For calcium, outputs ranged from 0.5 to 100 t km^{-2} yr^{-1}. Clearly a knowledge of rock and mineral weathering rates is relevant to water quality too. This important topic is dealt with at length in Chapter 6.

Conceptually, the reasons for different minerals exhibiting diverse rates of dissolution are straightforward. Where the lattice energy of the primary mineral is high compared with the bonding energy associated with the reaction products, which may or may not include a secondary mineral weathering product, transformation will be energetically favourable. The equilibrium solution solute concentration of weathering-derived species will be low, and the weathering reaction slow on a geological timescale. The actual rate will depend upon climate, site topography and all the other factors of soil formation.

Over recent decades, numerous geologists have attempted to draw up lists of minerals in order of 'weatherability', a so-called weathering sequence. One of the most useful in the context of soil chemistry is that of Jackson and Sherman (1953). It allows for considerable flexibility in interpretation, but once this is accepted it allows explanation of most field observations. The 13 stages are summarised in Table 2.2, which is based on the account in Sposito (1989). Many (but not all) of the minerals towards the resistant end of the sequence are secondary minerals that are the reaction products of primary mineral weathering.

Silicates

Simple tetrahedra

The great abundance of silicon in the Earth's crust means that silicates are of major importance in soil chemistry, a fact which will already be apparent from the wide occurrence of silicon in the idealised mineral formulae included in earlier sections. The fundamental unit which dominates the structures of silicate minerals is the SiO_4 tetrahedron. The silicon atom at the centre of each tetrahedron is bound via sp^3 hybrid orbitals to four oxygen atoms. The resultant anion for a single SiO_4 tetrahedron would have four negative charges for the oxygen atoms to have its preferred $1s^2$, $2s^2$, $2p^6$ electron configuration. The olivine group of minerals contains separate SiO_4^{4-} anions bound to divalent cations such as Mg^{2+}, Fe^{2+}, Mn^{2+}, Ca^{2+}, etc., to give a general formula M_2SiO_4. The lattice energy of such structures is not as great as when SiO_4 tetrahedra are joined to give chains, rings or sheets. Therefore the olivines are relatively readily weathered in soil (see the weathering sequence in Table 2.2).

Chain silicates

If SiO_4 tetrahedra join together to form chains in which adjacent tetrahedra share a common oxygen atom, a number of macromolecular structures with multiple charges, or, more strictly, free valences, become possible. The simplest structures are continuous chains, as shown in Fig. 2.4. In the figure the smaller Si atoms are shown (as open circles), although for a three-dimensional model they would be hidden by negatively charged O atoms rising vertically from the page. In the

Table 2.2. *Weathering sequence (after Jackson and Sherman (1953) and Sposito (1989)) for fine-grained minerals in soils*

Position[a]	Minerals	Comments
1	Gypsum, halite, other simple salts	Simple salts
2	Calcite, dolomite, apatite, aragonite	
3	Olivine, pyroxenes, diopside, hornblende etc.	Orthosilicates, chain silicates
4	Biotite, glauconite, magnesium chlorite, antigorite, nontronite	Layer silicates
5	Albite, anorthite, stilbite, microcline, orthoclase, etc.	Hard, feldspar framework silicates
6	Quartz, crystobalite, etc.,	SiO frameworks
7	Muscovite, etc.	Layers bound together by potassium ions
8	Interstratified 2:1 layer silicates and vermiculite	Secondary clay minerals
9	Montmorillonite, beidellite, saponite, etc.	Varieties of montmorillonite
10	Kaolinite, halloysite, etc.	
11	Gibbsite, boehmite, allophane, etc.	Hydrated aluminium oxides, etc.
12	Hematite, goethite, limonite, etc.	Hydrated iron oxides, etc.
13	Anatase, zircon, rutile, limenite corundum, etc.	Stable oxide frameworks

[a] 1, least inert; 13, most inert.

simplest chain, the pyroxene chain (Fig. 2.4(a)), the unit which is repeated to make up the chain is SiO_3^{2-}. Thus the pyroxenes are a group of chemically and physically related minerals which include enstatite and augite, mentioned earlier, hypersthene ($(Mg,Fe)O.SiO_2$), diopside ($CaO.MgO.2SiO_2$) and several other minerals (Fitzpatrick, 1980). Pyroxenes are only marginally more stable than silicates with single tetrahedra such as the olivine group.

The amphibole chain (Fig. 2.4(b)) is slightly more complex. The repeating subunit in this case is $Si_4O_{11}^{6-}$. The amphibole group includes hornblende and the tremolite–actinolite series.

Sheet silicates

A logical extension of the amphibole chain is a structure consisting of a network of SiO_4 tetrahedra linked to form hexagonal rings, and extending in two directions to form a continuous sheet. This structure is shown in Fig. 2.5. The repeating subunit in this instance is $Si_2O_5^{2-}$. The silicon atoms all lie in the same plane, but there are two planes of oxygen atoms, one containing the shared atoms, and the other the unshared, apical atoms. In minerals the silicate sheet occurs in

Fig. 2.4 Schematic representation of the linking of SiO_4 tetrahedra to form pyroxene chains (a) or amphiboles (b). The broken lines enclose the repeating units which make up the chains, SiO_3^{2-} and $Si_4O_{11}^{6-}$, respectively.

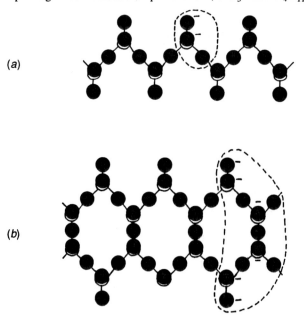

combination with sheets of octahedrally coordinated aluminium or magnesium atoms bonded to six shared or unshared oxygen atoms or hydroxyls.

Framework silicates

The framework silicates are composed of interlinked silicon tetrahedra extending in a three-dimensional array. Silica, the simplest member of this group, therefore has the structure $(SiO_2)_n$, and is placed well down the weathering sequence. In the feldspars, there is a large amount of isomorphous replacement of aluminium atoms for silicon atoms, leaving a negative charge which must be balanced by appropriate cations. If, for example, we take $(SiO_2)_4$ as a structural subunit in which one silicon atom in four is replaced by an aluminium atom, and the surplus charge is balanced by sodium, we finish up with a subunit of $NaAlSi_3O_8$. This is the unit formula of an alkali feldspar, albite. In practice, some of the surplus negative charge may be balanced by potassium, and a continuous series of alkali feldspars exists between the two extremes, albite and the potassium feldspar, $KAlSi_3O_8$, orthoclase.

Fig. 2.5 Schematic representation of the linking of SiO_4 tetrahedra to form silicate sheets. The repeating subunit is $Si_2O_5^{2-}$.

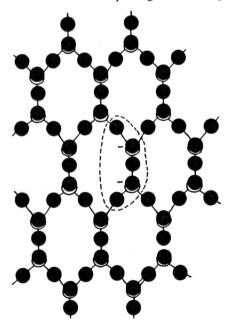

If the degree of isomorphous substitution is twice as great as described above, and the surplus negative charge is balanced by calcium, the resulting subunit formula becomes $CaAl_2Si_2O_8$, anorthite. There is a continuous series of feldspars, the plagioclase feldspars, between albite and anorthite, with steadily increasing amounts of calcium.

Other important sheet structures

In $Al(OH)_3$ (gibbsite) sheets, the aluminium is octahedrally coordinated (Fig. 2.6), and the octahedra are linked along their edges so that the aluminium atoms are in a hexagonal distribution throughout the same plane. There are two planes of hydroxide ions in the sheet. Because of the ratio of hydroxyl to aluminium, only two-thirds of the available octahedral positions between the hydroxyl planes are occupied by aluminium atoms. Such a sheet is known as dioctahedral (Greenland and Hayes, 1978).

The magnesium hydroxide $(Mg(OH)_2)$ sheet resembles the aluminium hydroxide sheet in that the magnesium is octahedrally coordinated, but because the ratio of hydroxyl to magnesium is 2 rather than 3, all of the available octahedral sites in the brucite $(Mg(OH)_2)$ sheet are occupied by a magnesium atom. Such a sheet is known as trioctahedral.

The minerals brucite and gibbsite contain stacked octahedral sheets of magnesium hydroxide or aluminium hydroxide respectively. Other important minerals consist of various combinations of silica tetrahedral sheets and brucite and gibbsite sheets. Often, however, many of the aluminium or magnesium atoms are replaced by atoms of other elements.

Fig. 2.6 Schematic representation of octahedrally coordinated aluminium.

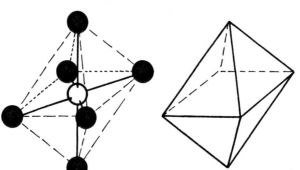

Isomorphous substitution

Although the ionic radius of aluminium (0.050 nm) is marginally greater than that of silicon (0.041 nm), it is possible for Al^{3+} to replace some of the Si^{4+} in tetrahedral sheets. Similarly Fe^{3+} (0.064 nm), Mg^{2+} (0.065 nm), Zn^{2+} (0.070 nm) and Fe^{2+} (0.075 nm) may replace the Al^{3+} in octahedral sheets (Brady, 1990). Replacement of an ion by another of similar size during mineral formation is known as isomorphous substitution. Isomorphous substitution will change the lattice energy, and hence the weatherability, of a mineral. Whenever the replacing ion has a lower positive charge than the ion it replaces, the mineral lattice is left with a net negative charge. The greater the degree of substitution, the greater the negative charge. It will be seen in Chapters 4 and 5 that this negative charge is a very important property of soils.

2:1 and 1:1 type minerals

Pyrophyllite contains a gibbsite sheet sandwiched between two silicon tetrahedral sheets, and is known therefore as a 2:1 type mineral. In both of the silicon tetrahedral sheets, the apical oxygen atoms are directed towards the central gibbsite sheet, and are linked to the sheet by replacing two-thirds of the gibbsite sheet hydroxyls. In terms of a structural subunit we therefore should start with two $Si_2O_5^{2-}$ units and an $Al_2(OH)_6$ unit. Four of the hydroxyls are then replaced by four oxygen atoms with free valences to provide a balanced structural subunit, $4[Al_2Si_4O_{10}(OH)_2]$. Pyrophyllite is relatively uncommon in soils, since it is formed by hydrothermal alteration of feldspars (Brown, Newman, Rayner & Weir, 1978).

Talc has a rather similar structure to pyrophyllite, but with a central brucite sheet in place of the gibbsite sheet. It is thus a hydrous magnesium silicate with a subunit $4[Mg_3Si_4O_{10}(OH)_2]$. The two aluminium atoms of pyrophyllite are replaced by three magnesium atoms in talc, i.e. all the octahedral sites in the brucite sheet are occupied by magnesium atoms, as mentioned earlier. Talc is also relatively uncommon in soils, since it is a high temperature alteration product of ultrabasic rocks or is formed by thermal metamorphism of siliceous dolomites (Brown *et al.*, 1978).

Muscovite is structurally similar to pyrophyllite except that one out of every four silicon atoms in the silicon tetrahedral sheets is replaced by an aluminium atom. This substitution leaves a surplus of negative charge which is balanced by potassium cations between the successive layers. These potassium ions hold adjacent layers permanently together. Thus the structural subunit of muscovite is $4[KAl_3Si_3O_{10}(OH)_2]$. It is a very common primary mineral.

Biotite is structurally analogous to talc, but again has one out of four silicon atoms in the silicon tetrahedral sheets replaced by an aluminium atom, creating a negative charge which is balanced by potassium cations which hold adjacent layers together. In addition, about one magnesium atom in three in the brucite sheet is replaced by iron(II). This does not change the charge distribution, but imparts a characteristic dark brown colour to the mineral. Thus the structural subunit of biotite is $4[K(Mg,Fe)_3AlSi_3O_{10}(OH)_2]$. Like muscovite, it is very common. In these micas, iron(II) may replace some of the magnesium, and iron(III) some of the aluminium, although replacement is more variable in the trioctahedral micas.

The primary minerals discussed in this section so far are all 2:1 type minerals. In some important minerals the layers consist of one silicon tetrahedral sheet and one gibbsite sheet, again held together by apical oxygens in the silicon tetrahedral sheet replacing some of the hydroxyls of the gibbsite sheet. Such 1:1 type minerals are discussed in the section on clay minerals.

Lattices containing sheet structures such as those described above tend to be somewhat more stable than the simple pyroxene chain type minerals (see Table 2.2). Silicates in which the macromolecular structure extends more or less uniformly in all directions to produce a three-dimensional network (the framework silicates discussed earlier) tend to have an even more stable lattice than many of the sheet silicates.

Important minerals other than silicates

In addition to the silicates, there are a number of other framework minerals which are relevant in soil chemistry and which will be mentioned elsewhere in this text from time to time. Some of them form very stable lattice structures and are therefore very inert. Most important are the oxides and hydrated oxides of iron and aluminium and, to a lesser extent, titanium. Apatites only generally occur in very small amounts, but are crucial to the healthy growth of plants because they are the sole source of phosphorus during early stages of soil formation.

Clay minerals

Kaolinite is a 1:1 type mineral with layers containing one silicon tetrahedral sheet and one gibbsite sheet, but with the apical oxygens of the silicon sheet replacing hydroxyls of the gibbsite sheet. It has a structural subunit formula $2[Al_2Si_2O_5(OH)_4]$, which is obtained by combining two $Si_2O_5^{2-}$ units with two $Al_2(OH)_6$ units with the loss of two hydroxyls from each $Al_2(OH)_6$ unit. The outer hydroxyls of the gibbsite sheet are strongly hydrogen bonded to the outer oxygens of the silicon tetrahedral sheet in the adjacent layer, giving a rigid, non-expanding structure of high stability (see Table 2.2). Kaolinite is very widespread in occurrence, since it is a stable decomposition product of weathering of granite or gneiss. Halloysite is closely related to kaolinite structurally, but the adjacent layers are separated by interlayer water which is fixed in position by hydrogen bonding. It is readily susceptible to dehydration to give a mineral with a kaolinite type structure. In his book *Soils*, Fitzpatrick (1980) includes excellent photomicrographs of the two minerals.

Serpentine is relatively a much more scarce 1:1 type mineral, each layer consisting of a silicon tetrahedral sheet and a trioctahedral brucite sheet. It has a structural subunit formula $2[Mg_3Si_2O_5(OH)_4]$. Details of other less common 1:1 type clay minerals may be found in the useful chapter by Brown *et al.* (1978).

Montmorillonite is a 2:1 mineral. The gibbsite sheet is sandwiched between two silicon tetrahedral sheets, as in pyrophyllite, and similarly apical silicon tetrahedral oxygen atoms replace two-thirds of the gibbsite hydroxyls. However, the structure differs from that of pyrophyllite in that *ca.* one aluminium atom in six is replaced by a magnesium atom. This creates an excess of negative charge which is satisfied by the presence of base cations in the interlayer spacing, which also serve to hold adjacent layers together. The mineral differs from kaolinite in that water is readily adsorbed into the interlayer spacing. This means that montmorillonite expands when wet and contracts when dry, and that the interlayer cations are relatively more labile, so that they may be displaced by excess of another suitable cation in solution. Such cations are said to be exchangeable. The mineral particles are usually very small and are also very soft as a consequence of the nature of the interlayer bonding. Montmorillonite is the simplest of a closely related series of clay minerals of similar structure now known as the smectites. All are 2:1 type silicates with variable cation exchange properties and basal spacing.

Hydrous micas are 2:1 minerals that are structurally virtually identical to the trioctahedral biotite or the dioctahedral muscovite that have been described earlier. The layers are linked by potassium, but the interlayers generally contain more water and less potassium per unit mass than the primary mineral micas. Although the minerals are hydrated, the interlayer bonding is sufficiently strong to prevent significant swelling and shrinkage.

Clay minerals of the vermiculite group are all 2:1 type minerals, but the central layer is variable in composition, containing aluminium, iron and magnesium in variable ratios. Both dioctahedral and trioctahedral vermiculites are found. Structurally they are similar to the hydrous micas, but with calcium or magnesium in the interlayer spacings rather than potassium. The alkaline earth elements cause different bonding between the layers, and consequently the minerals are capable of expansion when wet and contraction when dry. This may have a substantial effect upon structural properties of soils and their ease of cultivation. The occurrence of vermiculite is widespread, since it is an alteration product of basic or ultrabasic rocks, and is also found in other rock types (Brown *et al.*, 1978). As was the case for montmorillonite, the interlayer cations are exchangeable.

The chlorite group of minerals, which derive their name from the fact that they are frequently greenish in colour, have a very variable structure. A reasonable approximation of the structure of chlorite is two 2:1 type mica layers separated by a brucite layer. The negative charge on the mica layers is balanced by positive charge on the trioctahedral interlayer. The latter charge arises as a consequence of substitution of iron(III) and aluminium for magnesium in the brucite layer. Precise elucidation of the structures of chlorites formed in soils has proved rather difficult because of the variability of the mineral.

Interstratified minerals

The structure described above for chlorite might suggest that other mixtures of layers could occur in soils. Research over recent years has shown that this is indeed the case, and mixed-layer clays are now thought to be quite common. The resultant minerals are known as interstratified minerals (Brown *et al.*, 1978; Fitzpatrick, 1980).

Allophane

Although allophane is generally considered along with the clay minerals, its structure is highly variable and poorly defined. Probably both amorphous silica and aluminium hydroxide have been described as allophane at various times.

Imogolite

Unlike allophane, imogolite is now accepted to be a crystalline mineral with a definite structural form. It was first recognised in soils derived from glassy volcanic ash in Japan, but its occurrence in soils is now generally accepted to be more widespread than at first thought.

Soil minerals and soil chemical reactions

As stressed in Chapter 1, soil is biologically and chemically active, the organic and mineral components being intimately mixed in most soil types. Before considering the nature and driving forces behind chemical reactions which occur in soils, therefore, it is necessary to consider the nature and significance of soil organic matter. This is done in Chapter 3. The chemical reactions which take place in soil are then considered in Chapter 4.

3

Soil organic matter

Introduction

Although the total global reservoir of carbon is massive (more than 10^{19} kg), only a very small fraction of this carbon is actively involved in the fluxes of the carbon cycle, most of the Earth's carbon being locked away in sediments, in carbonate in oceans and in igneous rocks and fossil fuels. The Earth's active carbon pool consists of carbon in living organisms, carbon in the atmosphere and carbon in soil organic matter. It has been estimated that the mass of carbon in soil organic matter amounts to about 3×10^{15} kg and this is approximately five times the size of the atmospheric pool (Bohn, 1976). The latter pool is approximately the same size as that in living organisms.

Components of soil organic matter

Soil organic matter can, initially, be largely sub-divided into three main fractions:

(a) The decomposing residues of plants (as well as animals and microbes).
(b) The living soil biota – microorganisms, animals and plant roots.
(c) Resistant organic matter – chemically/physically protected.

Under most circumstances, more than 90% of soil organic matter is in resistant forms, although there is a continual flow of carbon from one pool to another. Figure 3.1 illustrates how this continual 'processing' of soil organic matter occurs. Generally, the system is in a state of dynamic equilibrium, with the organic matter pools remaining more or less constant with time.

Components of soil organic matter

Decomposing residues of plants

By far the most important input to soil organic matter is in the form of plant residues. These residues consist of litter, branches, root detritus and exudates. Figure 3.2 shows the main components of plant residues with their approximate relative proportions. Plant residues can provide an input to soil organic matter typically ranging from 11 tonnes carbon ha^{-1} yr^{-1} for tropical rain forests, to 6 tonnes for temperate forests, 3 tonnes for temperate grasslands, down to about 0.05 tonnes for deserts (Whittaker, 1975; Bolin, Degens, Kempe & Ketner, 1979). Of this plant carbon, between 60 and 70% is derived from the root system (Fogel, 1983), or, as it is often termed, 'rhizo-deposition'. This includes both soluble (amino acids, organic acids, carbohydrates) and insoluble (sloughed-off cells, etc.) material.

When plant residues enter the soil, there is an initial flush of decomposition (two-thirds of most plant residues entering soil decompose in one year), followed by a very much slower, steady breakdown.

Fig. 3.1 Simple schematic model demonstrating the continual decomposition of soil organic matter and its composite fractions. DPM = decomposable plant material; RPM = resistant plant material; BIO = microbial biomass; POM = physically protected organic matter; COM = chemically protected organic matter.

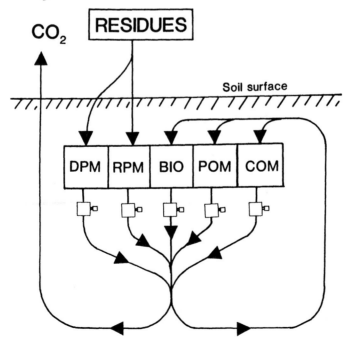

Soil organic matter

This pattern is largely due to two factors. The first is that some components of plant residues are much more readily decomposed than others. The second is that the formation of stable humic substances, after the initial flush of decomposition, prevents further rapid microbial attack.

To understand fully the origin and nature of soil organic matter, it is necessary to introduce the chemistry of decomposition of the most common components of plant residues and then to consider the formation and chemistry of the more resistant forms of soil organic matter, particularly the humic materials, which are the result of a tremendous range of chemical, as well as biological, transformations of plant (as well as animal and microbial) residues.

The ultimate end product of all microbial decomposition of organic matter in soil is carbon dioxide, assuming the soil environment is reasonably aerated. Because of this, carbon dioxide production is often used as an indicator of decomposition rates. This may not always be reliable as carbon dioxide can be produced in other ways in soil, such as through root respiration. One way around this problem is to use radio-labelled (^{14}C) organic substrates so that $^{14}CO_2$ production can be related to the individual substrate. Also, not all of the carbon from substrates produces carbon dioxide and a great deal more should be known about the fate of microbially-processed carbon before evolution

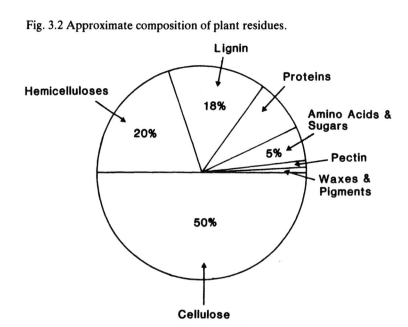

Fig. 3.2 Approximate composition of plant residues.

Components of soil organic matter

of carbon dioxide is used as a sole measure of substrate decomposition in the soil.

Cellulose

Cellulose frequently accounts for more than half of plant residue carbon (Fig. 3.2) as well as being a major constituent of fungal and algal cell walls. Cellulose occurs in a semi-crystalline state in these residues with a molecular mass of approximately one million. As with many large organic polymers entering the soil, the initial depolymerisation is catalysed by extracellular enzymes which are both free in the soil solution and adsorbed to soil colloids.

Depolymerisation of cellulose is carried out by specialised microorganisms in the soil, particularly fungi such as species of *Trichoderma*, *Fusarium* and *Aspergillus*, although bacteria (e.g. species of *Bacillus* and *Pseudomonas*) are also involved. Under aerobic conditions cellulose can be decomposed through to carbon dioxide, although under anaerobic conditions organic acids such as acetic acid are often generated:

$$C_6H_{12}O_6 \begin{cases} \xrightarrow{aerobic} + 6O_2 \rightarrow 6CO_2 + 6H_2O + energy \\ \xrightarrow{anaerobic} 2CH_3CH_2OH + 2CO_2 + energy \end{cases}$$

Cellulose is a simple polymer with glucose, as the only repeating unit, joined together by 1–4 linkages. Cellulose decomposition is catalysed by a number or suite of enzymes collectively referred to as cellulase. The activity of the cellulase enzyme system involves the loss of the crystalline structure of cellulose (referred to as 'conditioning' of the molecule) as well as subsequent depolymerisation and then oxidation of the individual glucose units.

During the depolymerisation of cellulose, and in fact during the catabolism of almost all organic polymers in soil, the molecule is attacked by two different types of depolymerising activity: 'endopolymerases' cleave the polymer at random, while 'exopolymerases' cleave only one or two units at a time from the non-reducing terminus of the polymer. After depolymerisation, the resulting short chains of glucose units are often referred to by specific names (two glucose units – 'cellobiose', three units – 'cellotriose') and the enzymes involved in their hydrolysis to single glucose units are referred to as cellobiase and cellotriase respectively.

The decomposition of cellulose is broadly representative of that of many other organic polymers in the soil, with initial depolymerisation by specialised microbial enzyme systems, but with subsequent release of simpler units which serve as a substrate for a much wider group of soil microorganisms and involve simple, non-specialised enzymes. Ultimately, the release of organic compounds such as glucose into the soluble carbon pool of the soil provides a substrate for all heterotrophic soil microorganisms possessing the simple glucose permease enzyme. The rate limiting step to the breakdown of polymers such as cellulose is invariably the initial depolymerisation, the flow of substrate carbon being extremely rapid when short units become available.

Cellulase activity, because of the widespread occurrence of the substrate, is found throughout the soil system – even in the guts of soil animals, although it is often more likely that it is the microorganisms in the gut that are producing the cellulase enzymes, rather than the soil animals themselves. The decomposition of wood in forest soils also involves considerable cellulolytic activity – cellulose often makes up half of the dry weight of woody tissues and occurs in filamentous microfibrils in a matrix of hemicelluloses and is sheathed in lignin. Most of the cellulolytic activity in forest soils is caused by specialised fungi known as brown rot fungi. The activity of these fungi is relatively rapid, leaving the more resistant lignin to be slowly decomposed by white rot fungi.

Hemicelluloses

The hemicelluloses are polymers of hexoses, pentoses and uronic acids and can often constitute up to one third of plant residue carbon.

A uronic acid, D-glucuronic acid

The hemicellulose pectin is a polymer of galacturonic acid with a

Components of soil organic matter

molecular mass of about 400 000. Although only a small part of plant dry weight, it has an important structural role, forming the mid-lamella region of plant cell walls.

The decomposition of pectin in soil is catalysed by the pectinase enzyme suite and occurs in three stages:

(a) In the first stage, extracellular pectin esterase enzymes catalyse the conversion of pectin to pectic acid:

$$(RCOOCH_3)_n + nH_2O \rightarrow (RCOOH)_n + nCH_3OH$$

(b) In the second stage, the molecule is depolymerised by extracellular exo- and endo-enzymes to produce short units of galacturonic acid:

<chemical structure of pectin showing repeating sugar units with COOH, OH, and COCH₃ groups>

carboxyl groups de-esterified in stage (a)

<chemical structure of galacturonic acid monomer with COOH and OH groups, labelled C × n>

(c) In the final stage, galacturonic acid residues are oxidised through to carbon dioxide:

$$2C_6H_{10}O_6 + 11O_2 \xrightarrow[\text{oxidase}]{\text{galacturonic acid}} 12CO_2 + 10H_2O + \text{energy}$$

Pectinase activity in soil is repressed by high sugar levels in fresh plant residues. This repression is a key controlling factor in determining the nature of microbial succession in plant litter decomposition. Species of the soil bacteria *Pseudomonas* and *Arthrobacter* are strongly pectinolytic, as are a number of actinomycetes and fungi.

Lignin

Lignin is the most resistant component of plant residues entering the soil and is the third most abundant component of plant residues, after cellulose and hemicellulose.

Lignin is a complex, non-uniform polymer of aromatic nuclei where the basic building unit is a phenyl-propane type structure:

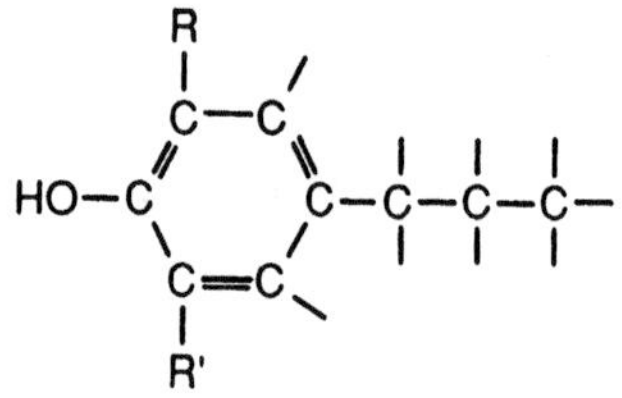

The R and R′ functional groups may be in three different forms where:

(1) R and R′ are H groups;
(2) R is H and R′ is OCH_3 (methoxyl);
(3) R and R′ are OCH_3 groups

The molecular mass of lignin is incredibly variable, ranging from tens of thousands to over a million. Different plant residues tend to contain lignin with different functional group characteristics. Coniferous trees, for example, tend to produce lignin nuclei where one group is hydrogen and the other methoxyl.

Lignin has a disorderly, cross-linked structure because of the considerable number of permutations involved in the bonding of repeating units. These units can be linked via strong ether bonds (C–O–C) or via even stronger carbon bonds (C–C). The links between units can be between phenolic rings, between propane side chains, or between a ring and a side chain.

A possible degradation pathway of coniferyl alcohol, which has a similar structure to the phenyl-propane subunit of lignin, is illustrated below:

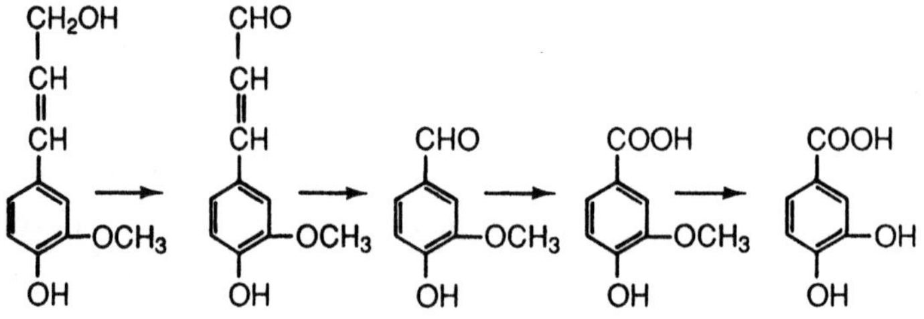

Coniferyl alcohol → Coniferaldehye → Vanillin → Vanillic acid → Protocatechuic acid

The 'random' structure of lignin and the strong linkages between subunits make lignin very resistant to the processes of microbial decomposition. Because of this, the proportion of carbon in the form of lignin in plant residues tends to increase with time in the soil.

Lignin decomposition in soil can be considered in three stages. In the first stage, the exposed hydroxyl groups are esterified. The second stage involves the depolymerisation process, while stage three involves the splitting of the phenyl ring after initial side chain removal.

While many soil microorganisms can degrade some of the components of the lignin molecule, only a very limited number can degrade the entire lignin structure. Most of these microorganisms are fungi and are collectively termed 'white rot' fungi. White rot fungi include species of *Phanaerochaete* and *Coriolus*. These fungi must be regarded as amongst the most specialised of decomposers, producing a vast array of enzymes including esterases, phenolases, and peroxidases for relatively small energy returns.

Lignin degradation in soil can be inhibited by high concentrations of nitrogen in the soil. Because of this, it has been suggested that lignin degradation may be a secondary metabolic function induced by nitrogen starvation rather than a function of primary metabolism (Kirk & Fenn, 1982). It is possible that the introduction of radiocarbon-labelled analogues of lignin (Kirk *et al.*, 1975) will enable the complex degradation of lignin in soil to be effectively characterised. One thing for certain, however, is that, despite the complexity of lignin degradation, after depolymerisation it becomes a substrate/energy source for an increasing number of soil microorganisms. Ultimately, of course, much of the carbon found in lignin will be lost from the soil as carbon dioxide.

Soil biota and organic matter turnover

This section will attempt not to characterise fully the living soil biota, but briefly to explain its role as the driving force in the turnover of organic matter in soil.

All organic matter in the soil, whether fresh or partially decomposed plant/animal/microbial residues or more resistant forms of organic matter, will eventually be processed by the soil microbial component of the soil biota. The microbial biomass, therefore, is the 'motor' which turns the cycling not just of carbon, but of the 'biological' elements such as nitrogen, phosphorus and sulphur. These elements form the main building blocks of cellular tissue and are initially absorbed by the lower forms of life as simple inorganic forms which are then converted to organic constituents within the cell. The death and subsequent decay of living tissues releases inorganic ions and, so, the cycle continues.

The soil decomposer community and microbial succession

Clearly, many of the nutrients in the life cycle of plants, animals and microorganisms are cycled in the soil between the soil organic matter and the inorganic nutrient pool. Although detailed characterisation of the soil community which carries out this cycling is beyond the scope of this book, it is important to note that it does not simply involve the activity of the primary decomposing microorganisms (fungi and bacteria), but also incorporates the activity of the soil animals (protozoa and larger soil animals such as earthworms, mites, springtails etc.) and of the plants, since they represent the biggest input of carbon and organically bound nutrients.

Biological processing of organic matter in soil often involves initial comminution of debris by soil animals as well as microbial attack. The latter involves the depolymerisation of polymeric components by extracellular exo- and endo-enzymes and the intracellular processing of depolymerised subunits as well as smaller organic molecules. This sequence in the decomposition of complex substrates suggests that not all of the microbial community is equally involved in all stages of attack. In fact, initial depolymerisation of recalcitrant fractions such as lignin tends to be the domain of a few specialised microorganisms, although the subsequent breakdown products provide a substrate for an increasingly wide decomposer community. A clear account of this type of 'substrate succession' for saprophytic fungi is provided by Cooke and Rayner (1984). In addition to substrate succession of decomposer organisms in soils, one can classify decomposer microorganisms on the basis of whether they are 'autochthonous' or 'zymogenous'. The activity of the zymogenous microorganisms is associated with the flush of decomposition which occurs when fresh substrate/residue enters the soil, while autochthonous microorganisms represent the indigenous soil population which 'ticks over' when concentrations of available carbon are low. In terms of Michaelis–Menten kinetics (Fig. 3.3), the zymogenous decomposer population tends to have a high maximum specific growth rate (μ_{max}), but a relatively low Michaelis constant (K_m) while the autochthonous decomposer population tends to have a low μ_{max}, but a relatively high value for K_m.

Ultimately, after the successional flush of the zymogenous decomposer community, the processed carbon from plant or other residues

Soil biota and organic matter turnover

will be lost from the soil as carbon dioxide and the remainder incorporated into resistant soil organic matter.

Kinetics of organic matter breakdown

The rate of processing of organic matter by the soil microbial biomass depends upon a number of environmental as well as substrate-related factors. In general, at a given time, organic matter processing is in a state of dynamic equilibrium which can be affected by changes in environmental controls such as soil water potential, pH, temperature, aeration etc. or by an addition of a decomposable organic substrate such as is provided by fresh plant residues. The rate of organic matter decomposition is generally considered to follow first order kinetics, i.e.:

$$dS/dt = kS_0 \tag{3.1}$$

where S is the substrate concentration at time t, S_0 is the initial substrate concentration and k is the first order decay constant. Integrating equation (3.1) with respect to S gives:

$$S = S_0 e^{-kt} \tag{3.2}$$

Fig. 3.3 Comparative Michaelis–Menten kinetics of autochthonous and zymogenous decomposer populations.

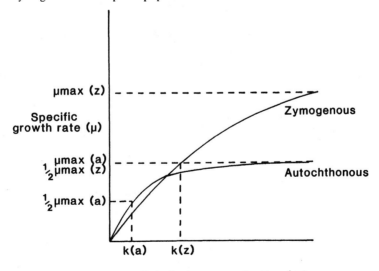

$$\mu = \frac{\mu max.S}{k+S}$$

Thus $\ln(S/S_0) = -kt$ and $\log(S/S_0) = -kt/2.303 = k't$. The relationship between substrate concentration and time can be represented graphically as in Fig. 3.4.

The turnover time of a substrate decomposing according to first order kinetics is defined as the time taken for 90% decomposition. Thus, when t = turnover time, $S/S_0 = 10/100 = 0.1$, and $\log(S/S_0) = -1$. If $k't = -1$, $t = 1/k'$, and thus turnover time is equal to the reciprocal of the rate constant (i.e. $1/k'$). Readily decomposable substrates such as glucose ($k' = 0.1$–0.3 day^{-1}) and cellulose ($k' = 0.05$–0.1 day^{-1}) have a very short turnover time in the soil (glucose – around 3–10 days; cellulose – around 10–25 days), while resistant plant residue components such as lignin ($k' = 0.002$–0.003 day^{-1}) have considerable residence times (measured in years rather than days). Chemically protected humic components of resistant organic matter will decompose even more slowly ($k' = 0.0002$–0.0004 day^{-1}) than lignin. Pools of organic matter such as this in the soil are continually being decomposed and continually being renewed by fresh carbon flow (Fig. 3.1). The time taken to decompose this type of material is therefore a difficult concept and we often refer to mean residence time (still equivalent to $1/k'$) or mean turnover time which can be estimated by ^{14}C dating of specific organic matter fractions in the soil. More detailed discussion of how these chemical fractions can be isolated is presented later.

Resistant soil organic matter

Experiments with ^{14}C-labelled plant residues have demonstrated that, while much of the carbon in these residues is lost to the atmosphere as carbon dioxide, a portion remains in the soil for many years. Label searches have demonstrated that this carbon is in microbial tissues and in organic matter which is highly resistant to decomposition.

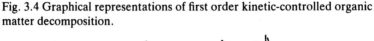

Fig. 3.4 Graphical representations of first order kinetic-controlled organic matter decomposition.

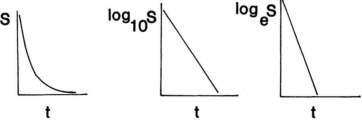

Incorporating ^{14}C-labelled ryegrass into soil, Jenkinson (1965) demonstrated that 10% of the carbon was present in microbial biomass after 1 year and this had dropped to 4% after 4 years. This flow of residue-derived carbon through the microbial biomass, and through the resistant organic matter fraction, is linked to other cycles and hence drives the flow of nitrogen, sulphur, and phosphorus in the soil.

Chemically protected organic matter

There is no detailed understanding of the structure of chemically resistant organic matter in soil (Swift, Heal & Anderson, 1979). The formation of this fraction of soil organic matter, often referred to as the humic component of soil, however, involves the degradation and demethylation of lignin polyphenol monomers and then the oxidation of these polyphenols to quinones which are subsequently condensed with low molecular mass microbial cell products to form either the humic molecules or their immediate precursors. The low molecular mass microbial products are mainly amino acids, and to a lesser extent, nucleic acids and phospholipids. This scheme of formation of humic material is referred to as the 'polyphenol theory', because the humic molecules are the result of condensation of phenolic material (Fig. 3.5). This scheme, proposed by Flaig & Sochtig (1964), contrasts markedly with the long-held view, proposed by Waksman (1936), that humic substances simply result from the incomplete degradation of lignin (the 'lignin theory'). This view was disproved when it was found that humic substances could be synthesised by fungi in the absence of lignin (Russell, Vaughan, Jones & Fraser, 1983).

Although the pool of humic carbon in the soil tends to be at a dynamic equilibrium, the actual condensation reactions forming the humic molecules are not enzymatically controlled and so there will be considerable variability in the exact nature of the molecules.

Figure 3.6 illustrates the main forms of carbon, and their typical relative proportions, in humus. Of course these proportions vary considerably from one soil to another. Hatcher, Maciel & Dennis (1981) demonstrated, using magic angle spinning, ^{13}C, nuclear magnetic resonance (NMR) that the aromatic carbon fraction can range from over 90% to less than 40%.

The humic components tend to make up the largest proportion of soil organic matter, reflecting resistance to decomposition. Jenkinson and Rayner (1977) assigned to this fraction a first order decay constant of 0.00035 yr^{-1} when modelling the turnover of organic matter in long-term cultivation experiments at Rothamsted Experimental Station in England.

Humic molecules are often bound to other solid components of the soil matrix. Clay–humate complexes are thought to form when humates are adsorbed to clay minerals by polyvalent cations such as Ca^{2+} and Fe^{3+}, and by association with hydrous oxides, either through coordination (i.e. ligand exchange) or through anion exchange via positive sites which exist on iron and aluminium oxides (Stevenson, 1982). These positive sites on sesquioxides will not exist at soil pH values above 8, and so clay–humate complexes are less likely to form in highly alkaline soils.

Not all chemically resistant organic matter in soil is in the form of complex aromatic carbon, large pools of potentially more degradable

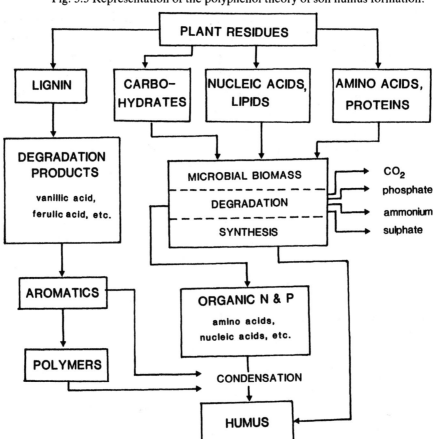

Fig. 3.5 Representation of the polyphenol theory of soil humus formation.

carbon being 'protected' from microbial attack by a layer of less palatable phenolics. Because of this, amino acid residues of great age have been extracted from soil (Wagner & Mutakar, 1968), no doubt protected by the kinds of complexes illustrated below.

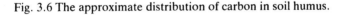

A quinone–protein complex

Physically protected organic matter

While most of the resistant organic matter in soil is in chemically stable forms, there appears to be an important fraction of soil organic matter that is protected physically in the soil. Evidence for this fraction comes from the considerable release of carbon dioxide (or mineral nitrogen) that occurs when soils are physically disrupted, either through soil grinding (Powlson, 1980), shearing and compression (Rovira and Greacen, 1957) or through various forms of cultivation in the field. Most of this physically protected organic matter is in the form of polysacchar-

Fig. 3.6 The approximate distribution of carbon in soil humus.

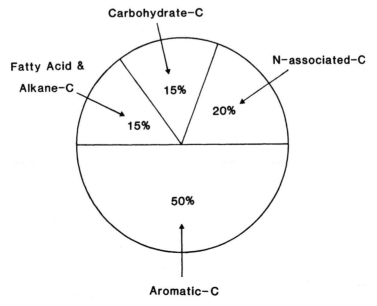

ides. Up to 10% of organic carbon in soils can be in this carbohydrate form (Cheshire, 1979).

Powlson (1980) found that grinding soil released carbon from both microbial biomass and from non-biomass organic matter. It appears from this study that grinding can kill about a quarter of microbial biomass and that this contributes between 25% and 50% of the additional carbon dioxide evolved when soils are ground. The extra non-biomass organic matter decomposing after grinding appears to represent about 0.5% of the total soil organic carbon, although no doubt this figure will vary depending on both the severity of physical disruption and on soil type.

Jenkinson & Rayner (1977) recognised the presence of physically protected organic matter when modelling turnover of carbon in soils, and assigned this fraction a first order rate constant of 0.014 yr^{-1} (mean residence time – 71 yr) compared to 0.00035 yr^{-1} (mean residence time – 2857 yr) for chemically protected organic matter.

The precise nature of physical protection of organic matter in soil is open to speculation and, in any case, probably varies with soil type. It is clear, however, that it is the physical disruption of otherwise stable aggregates that exposes protected organic matter to biological attack. It must be assumed that the connectivity of the pores containing organic matter in these aggregates is restricted (i.e. excluding microbial decomposers), otherwise protection would not be conferred. It is likely that clays play a crucial role in aggregate protection of organic matter and the possible mechanisms of this have been considered (Crasswell and Waring, 1972).

Detailed characterisation of soil organic matter

The earliest extraction of humic substances from soil was probably carried out by Achard in 1786. This was an alkali extraction of peat and, upon subsequent acidification of the extract, produced a dark, amorphous precipitate which has now become known as humus.

Most current procedures used in characterisation of soil organic matter are based on the extraction of soil with alkaline solutions, particularly sodium hydroxide. Sodium hydroxide is thought to extract about 80% of the soil's total organic matter (Stevenson, 1982). Figure 3.7 illustrates the steps involved in the extraction of humic substances from soil. The alkali extraction is often preceded by a mild acid treatment (0.1 M HCl) to improve the subsequent yield of alkali-soluble material. Acid pre-treatment yields some soluble, fulvic acid material.

Characterisation of soil organic matter

The relative proportions of fulvic acid and humic acid found by the sorts of steps outlined in Fig. 3.7 very much depend upon soil type (Linehan, 1978). The depth in the soil profile from which the samples were taken also influences these proportions (Stevenson, 1982).

Chemical fractions of soil humus

Having carried out the fundamental extraction procedures of Fig. 3.7, some further characterisation of the resulting fractions has been achieved through elemental analyses, infra-red spectroscopy, and through an array of gel chromatographic procedures. The validity of further chemical characterisation of extracted organic matter fraction must, however, be seriously questioned because of the drastic nature of the extraction procedures!

Humic acid

The most studied group of the humic substances is the humic acid fraction. Elemental analysis of humic acid has shown it to consist

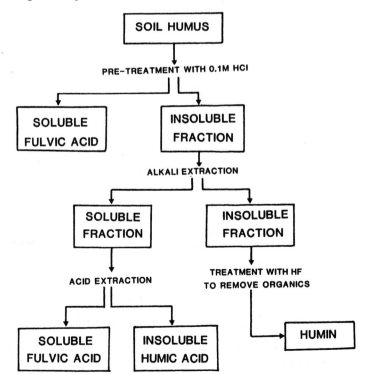

Fig. 3.7 Steps involved in the fractionation of soil humic substances.

largely of carbon and oxygen (about 50 and 40% respectively), but also to contain hydrogen (about 5%), nitrogen (about 3%), phosphorus and sulphur (both less than 1%). Elemental composition of humic acid has been reported, amongst others, by Russell *et al.* (1983) and Swift & Posner (1972). The structure of humic acid is polyphenolic with an array of functional groups including carboxyls, hydroxyls (aromatic and aliphatic), amides and carbonyls.

It is reasonable to assume that humic acid is a complex of closely related macromolecules. Gel filtration suggests that the molecules range from less than 1000 to more than 100 000 daltons (Swift & Posner, 1972), with the lower mass representing the younger material. The exact composition of humic acid varies from one source to another but, more importantly, varies with the nature of the extraction used.

Fulvic acid

The fulvic acid fraction of humus has not been as extensively studied as humic acid. Fulvic acid contains less carbon and more nitrogen and oxygen than humic acid and it seems most likely that, while its structure is broadly similar to that of humic acid, it has a smaller proportion of aromatic units and greater peripheral aliphatic chains (Schnitzer & Khan, 1978). Carboxyl functional groups appear to be more evident in fulvic acid than humic acid and are probably heavily substituted on the aliphatic chains (Anderson, 1982).

Humin

The alkali-insoluble humin fraction of soil humus is the least studied of the three fractions. This is probably partly due to the extreme difficulty in obtaining 'pure' humin which is free of inorganic impurities. Removal of these inorganic impurities involves drastic treatment of the fraction with hydrofluoric acid (see Fig. 3.7). The structure of humin appears to be broadly similar to that of humic acid, although it contains less nitrogen (about one third that found in humic acid). The insolubility of humin is more likely due to binding to mineral impurities rather than to fundamental structural differences between humin and the other soil humic fractions.

Soil organic matter and soil structure

Soil organic matter has a vital role to play in the establishment of soil structure and in the maintenance of stability of soil structure. It is largely organic matter that binds primary particles together to form aggregates sufficiently stable to withstand erosion by the wind, the impact of falling raindrops and the movement of machinery and animals over the soil surface.

Tisdall & Oades (1982) identified three different modes in which soil organic matter contributes to soil structure as a binding agent, namely transient, temporary and persistent modes. Each mode is considered briefly below.

Organic matter as a transient binding agent

The transient binding agents are mainly plant and microbial products and are readily decomposed in the soil, but are continually produced. It is the microaggregates (20–50 μm) that tend to be held together by this type of amorphous soil organic matter. Much of the evidence for this mechanism comes from the loss of aggregate stability where soils are treated with reagents, such as periodate and tetraborate, which destroy carbohydrate (Cheshire, Sparling & Mundie, 1983).

The adhesive properties of polysaccharides in contributing to aggregation in soils largely result from weak physical forces such as van der Waals forces and from hydrogen bonding, but are facilitated by the large surface area presented by clay particles in particular. Neutral and acidic polysaccharides appear to be able to bind particles into aggregates with the effectiveness of the acidic polysaccharides being controlled by the uronic acid content (Martin, 1971) and the effectiveness of the neutral polysaccharides being controlled by the molecular mass, the greater molecular mass enabling a greater number of linkages between the soil and the parent molecule.

A number of the neutral polysaccharides such as the dextrans involved in aggregate formation in soil are produced by algae. These organisms may, therefore, have a key role in soil aggregation, particularly in tropical and subtropical situations (Lynch, 1983) and production/application of algal inocula may provide a potential management strategy in some soils that are sensitive to wind and water erosion.

Organic matter as a temporary binding agent

The hyphae of filamentous fungi (including mycorrhizal fungi) and algae, as well as the fine roots of plants, make up this group of binding agents. They tend to accumulate in soil over a period of time and can persist for months or even years. The larger soil particles (particularly the sand-sized fraction) tend to be held together by these types of agents (Forster, 1981). The various microbial filaments tend to have sticky surfaces as a result of extracellular polysaccharide production, and the tips of the developing roots produce sticky polysaccharidic mucigel. These sticky substances enable the filaments to adhere strongly to the soil particles and facilitate the physical enmeshing process.

Organic matter as a persistent binding agent

The main persistent binding agents are the resistant humic constituents resulting from the decomposition of plant, animal and microbial residues. This group may often be associated with amorphous aluminium and iron in soils.

Organic matter and soil structural stability

Additions of fresh organic matter to soil, particularly as crop/plant residues, will contribute to the three organic binding modes just discussed. Continuous inputs of plant residues are therefore essential for the maintenance of soil structural stability. Greenland, Rimmer & Payne (1975) have suggested that non-calcareous soils with organic matter contents below 3.4% are likely to suffer structural deterioration whereas soils containing 4.3% or more organic matter are likely to be structurally stable. Continuous cultivation by modern, high input farming techniques tends to cause oxidation and loss of this vital organic matter component of soil structural stability. Strutt (1970) also suggested that organic matter concentrations of 3% or less may be satisfactory for prolonged cereal production on stable soils but will rapidly become inadequate for crops on unstable soils, particularly those high in sand or silt and low in clay.

Replacing lost organic matter is a slow process. A soil containing 3% organic matter contains approximately 75 tonnes of organic matter over 1 hectare to the plough depth (150 mm). To increase this to 4% organic matter would take close to 100 years if 2 tonnes per hectare of crop residues were added annually. This is a fairly typical rate of residue addition under an organic farming type regime. The slow accumulation of soil organic matter reserves is partly because some two-thirds of the residue carbon is decomposed to carbon dioxide and lost to the atmosphere.

Maintenance of soil structural stability through organic matter additions will be vital to future soil conservation, not just in the tropics where soils are particularly sensitive and rainfall is often erosive, but also in temperate regions where machinery in particular can damage structurally-weakened soils.

Rates of degradation and accumulation of organic matter in soil

Since the organic matter content of most soils is not steadily increasing, it is safe to assume that the continuous addition of organic matter as plant, animal and microbial residues is matched by processes of decomposition which lead to the return of organic carbon to the atmosphere as carbon dioxide. In fact the organic matter content of most soils can be said to be in a state of dynamic equilibrium between the processes of degradation and accumulation.

Turnover of soil organic matter

The process by which organic matter is continually decomposed and renewed is known as turnover. Jenkinson & Ladd (1983) showed that, for an arable soil under continuous wheat cultivation at Rothamsted in southern England, with 26 tonnes of soil organic carbon per hectare and an annual carbon input of 1.2 tonnes, the turnover of organic matter is rapid:

$$\text{Turnover (steady state)} = \frac{\text{soil organic carbon (t ha}^{-1}\text{)}}{\text{input (or output) of organic carbon (t ha}^{-1}\text{ yr}^{-1}\text{)}}$$

$$= \frac{26 \text{ t ha}^{-1}}{1.2 \text{ t ha}^{-1} \text{ yr}^{-1}} = 22 \text{ yr}$$

The organic matter of this soil is therefore turning over every 22 years. Turnover, however, does not just apply to the total organic matter content of soil, but also to individual fractions. Because these different fractions tend to decompose (and be renewed) at very different rates, then their rates of turnover will correspondingly differ.

Numerous attempts have been made to model this turnover of organic matter in soil mathematically, using the approach illustrated in Fig. 3.1. Perhaps the best example has been that of Jenkinson & Rayner (1977), who used data from part of the same long-term field experiment as that studied by Jenkinson & Ladd (1983) to establish a mathematical model predictive of the turnover of organic matter in soil and the ultimate equilibrium status of the complete soil organic matter pool and its component fractions. The model was constructed using data from soils where the input of crop residues had been more or less constant over the past 100 years or so. These data included long-term changes in total soil organic matter, rates of decomposition of ^{14}C-labelled residues in soil, radiocarbon dating of soil humus, the effect of thermonuclear radiocarbon on radio carbon age (the 'bomb' effect) and recent information on the size of the microbial biomass. The model separates soil organic matter into five components and assumes all decompose according to first order kinetics (Fig. 3.4) with the following rate constants and half lives

   (i) decomposable plant material (DPM)
       $K_{DPM} = 4.2 \text{ yr}^{-1}$ $t_{1/2DPM} = 0.165 \text{ yr}$
   (ii) resistant plant material (RPM)
       $K_{RPM} = 0.3 \text{ yr}^{-1}$ $t_{1/2RPM} = 2.31 \text{ yr}$
   (iii) microbial biomass (BIO)
       $K_{BIO} = 0.41 \text{ yr}^{-1}$ $t_{1/2BIO} = 1.69 \text{ yr}$
   (iv) physically protected organic matter (POM)
       $K_{POM} = 0.014 \text{ yr}^{-1}$ $t_{1/2POM} = 49.5 \text{ yr}$
   (v) chemically protected organic matter (COM)
       $K_{COM} = 0.00035 \text{ yr}^{-1}$ $t_{1/2COM} = 1980 \text{ yr}$

The model predicts that for an annual input of plant residue carbon of 1 tonne per hectare to the Rothamsted soil, the equilibrium status of the soil organic matter will be 12.2 t COM, 11.3 t POM, 0.28 t BIO, 0.47 t RPM and 0.01 t DPM. Clearly, because of the slow turnover of chemically and physically protected organic matter, these fractions predominate in most situations.

The model developed by Jenkinson and Rayner is progressively being updated so it can be adjusted to include variations in conditions of both soil and climate, although the basic principles still apply and it provides a useful understanding of the kinetics of organic matter turnover in soil.

Accumulation and loss of soil organic matter

From the discussion of turnover of soil organic matter, it is evident that any change in turnover will lead to either an increase or a decrease in soil organic matter reserves until a new equilibrium is reached. Such a change may come about either as a result of a change in soil conditions (e.g. a change in soil pH, aeration, temperature etc.) which affects the rate at which the microbial biomass processes the different soil organic matter fractions, or as a result of a change in the nature of the substrate entering the soil. The time taken to reach a new equilibrium status is measured in thousands of years, but because of the nature of first order kinetics, most of the change will take place in the first few hundred years. The magnitude of the gain or loss in soil organic matter reserves is generally small, unless there is a radical change to the rates of biological attack. As outlined earlier, a change of just 1% (of soil dry weight) in the organic matter content of a soil will require an increase or decrease in the annual plant residue additions to the soil of about 2 tonnes per hectare over a one hundred year period!

Of course, extreme situations of soil organic matter accumulation or loss can occur. The formation of peats is an example of where unfavourable conditions for decomposition due to poor drainage, and sometimes high acidity/reduced aeration, cause rapid accumulation of largely undecomposed or partially decomposed plant remains. Accumulation of litter and humus beneath a first rotation forest, particularly of conifers with slowly decomposing litter, provides another example of relatively rapid organic matter accumulation. The humus layer of a podzol beneath conifers has been observed to increase by as much as 10 cm over a 40 year period (Billett, FitzPatrick & Cresser, 1988). This rapid accumulation is both the result of low pH (acidity generally retarding turnover of organic matter) and the high lignin content of the litter which reduces the substrate quality for the decomposer organisms.

Rapid loss of soil organic matter can arise when peat soils are drained and aeration is allowed to take place. The loss of much of the British fenland peats since drainage is an example of this and has been accelerated by the additions of lime and fertiliser needed to sustain crop productivity.

The organic chemistry of anaerobic soils

Although most arable soils are dominated by aerobic soil conditions, anaerobic conditions can prevail either temporarily in the bulk soil or, more commonly, in microsites. Of course, some non-arable soils, particularly in situations of poor drainage, may experience prolonged anaerobic conditions. There is an intimate link between anaerobiosis and organic matter dynamics in soil.

Soil organic matter and the generation of anaerobic conditions

The decomposition of organic matter in soils by heterotrophic organisms (i.e. organisms which require pre-formed organic carbon as a source of carbon and energy), including microorganisms and soil animals, generally involves the oxidation of reduced organic substrates under aerated conditions. The electrons removed in these oxidations are passed down an electron transport chain inside the decomposer organisms which enables energy from the electrons to be harnessed in adenosine triphosphate or ATP. The decomposers then use oxygen to accept the 'spent' electrons, producing water. Oxygen is referred to as the 'terminal electron acceptor'.

As oxygen is used in this way, it will generally be replaced by fresh oxygen from the atmosphere which diffuses through the soil to the sites of microbial respiration. The rate of this diffusion is governed by Fick's Law:

$$Q = D \frac{dc}{dz}$$

where Q is the diffusion rate of gas (kg m^{-2}), D is the diffusion coefficient in soil (m^2 s^{-1}), c is the concentration of gas in soil air (kg m^{-3}) and z is the depth in soil (m).

In well-aerated soils, oxygen concentrations are rarely reduced much below the 20% normally found in the atmosphere. This is because the diffusion coefficient of oxygen in air (D_{air}) is very rapid (2 × 10^{-5} m^2 s^{-1}). In soils with a high moisture status, however, oxygen replenishment becomes restricted because of the considerably lower diffusion coefficient of oxygen in water (0.2 × 10^{-8} m^2 s^{-1}).

Because of the importance of water in controlling oxygen diffusion in soils, it is the water potential of soil and the pore structure (small pores remain water-filled at lower water potentials than larger pores) that determine the rate at which oxygen demand is replenished in soil. In fact there is a continuum from soils that are flooded and continuously anaerobic, through aggregated, well drained soils which are periodically anaerobic (particularly in aggregate microsites – see the next section) to excessively well drained, light textured soils for which it is highly unlikely that anaerobiosis will occur.

The process of terminal electron acceptance by oxygen will continue down to millimolar concentrations of oxygen in the soil atmosphere and dissolved in the soil water. When respiratory demand causes oxygen to drop below this value, many decomposers cease to be active. These organisms, which can only use oxygen as their terminal electron acceptor, are referred to as 'obligate aerobes' and include all soil animals and soil fungi and a great many soil bacteria.

For some decomposer bacteria, nitrate (NO_3^-) may replace oxygen as the terminal electron acceptor, the nitrate reductase enzyme replacing cytochrome oxidase. These organisms, which can remain active under aerobic and anaerobic conditions, are referred to as 'facultative anaerobes'. They carry out the reduction of nitrate, ultimately to free nitrogen, a process known as denitrification. Once oxygen and nitrate have been largely used through respiratory demand, there follows a sequential reduction of possible electron acceptors. Generally, different bacterial decomposers have evolved to use these different acceptors. Bacteria that can only use electron acceptors other than oxygen are referred to 'obligate anaerobes' while, as previously stated, facultative anaerobes are able to switch between oxygen and alternative electron acceptors.

The electron acceptor in operation in soil at any one time is determined by the soil redox potential which is a measure of the likelihood of a substance to gain (reduction) or lose (oxidation) electrons, as discussed in Chapter 4.

The sequence of reduction of terminal electron acceptors in the soil environment is listed in Table 3.1. This shows the approximate redox potential at which each reduction couple operates. The first four couples illustrated are carried out by facultative anaerobes and the last

Table 3.1. *Sequence of redox couples and associated microbial processes operating in the soil environment, with associated redox potential at neutral pH*

Redox couple	Microbial process	Redox potential mV	Soil organisms involved
$O_2 \xrightarrow{+e^-} H_2O$	Aerobic respiration	+820	Plant roots, aerobic microbes, animals
$NO_3^- \xrightarrow{+e^-} N_2, N_2O$	Denitrification	+420	*Pseudomonas*,
$Mn^{4+} \xrightarrow{+e^-} Mn^{2+}$	Manganese reduction	+410	*Bacillus*, etc.
Organic $\xrightarrow{+e^-}$ organic matter acids	Fermentation	+400	*Clostridium*, etc.
$Fe^{3+} \xrightarrow{+e^-} Fe^{2+}$	Iron reduction	−180	*Pseudomonas*
$NO_3^- \xrightarrow{+e^-} NH_4^+$	Dissimilatory nitrate reduction	−200	*Acromobacter*
$SO_4^{2-} \xrightarrow{+e^-} H_2S$	Sulphate reduction	−220	*Desulfovibrio*
$CO_2 \xrightarrow{+e^-} CH_4$	Methanogenesis	−240	*Methanobacterium*

two by obligate anaerobes. The facultative anaerobes are generally present in high numbers in most soils, often accounting for up to 10% of the total bacterial population (Skinner, 1975), while obligate anaerobes are thought to be present in relatively low numbers and are often found as resistant spore bodies, inactive until favourable physicochemical conditions prevail.

The top four redox couples illustrated in Table 3.1 may well occur in either soils with anaerobic microsites or in soils which are only periodically anaerobic.

The reducing power generating anaerobic conditions in soil generally comes from highly decomposable organic matter such as plant residues and animal slurries. It is this form of soil organic matter that supplies most of the available carbon for respiration, whether aerobic or anaerobic.

Anaerobic microsites in soil

Anaerobic microsites are most likely to develop in the centres of water-saturated soil aggregates. This is because the slow diffusion of oxygen into such sites is exceeded by the rate of consumption of oxygen by aggregate-dwelling microorganisms. Greenwood (1975) calculated the critical aggregate radius (assuming spherical aggregates) for the onset of anaerobic conditions at the centre of the aggregate for different respiratory oxygen demands at that centre. To make this calculation, a derivative of Fick's Law (the law governing rates of gaseous diffusion) was used:

$$a^2 = 6CD/R$$

where a is the aggregate radius (m), C is the difference in dissolved oxygen concentration between the aggregate's surface and centre (ml O_2 ml^{-1}), D is the diffusion coefficient of oxygen through the aggregate (m^2 s^{-1}) and R is the respiration rate inside the aggregate (ml O_2 m^{-3} s^{-1}). It was found that anaerobic conditions will occur in water-saturated soil aggregates of radius 10 mm or greater for typical rates of soil respiration. These conditions will most likely be found in well-aggregated soils in the spring when both moisture contents and biological activities are high. The fact that anaerobic processes can occur in most freely draining soils suggests that anaerobic microsites can readily develop in soil and that Greenwood's calculations may overestimate the critical radius for aggregates in which anaerobiosis occurs.

The organic chemistry of anaerobic soils

Competition between electron acceptors in anaerobic soils

It is a good general rule that the presence in soil of electron acceptors that are high in the order of redox couples of Table 3.1 tends to inhibit respiration involving an electron acceptor of a lower redox potential. For example, sulphate reduction will be inhibited by the presence of oxygen, nitrate, manganese as Mn(IV) and iron as Fe(III).

A high concentration of one particular electron acceptor in soil buffers the soil redox potential. A high concentration of nitrate in the soil solution, for example, maintains soil redox at around 224 mV (at pH 7), even though fresh organic matter may be providing considerable reducing power.

Consequences of anaerobiosis in soils

Anaerobiosis in soil directly affects plant growth and productivity, but also indirectly affects plant growth through microbially mediated changes in the cycling of organic matter and nutrients (e.g. denitrification will remove plant-available nitrate from the soil solution), through microbial production of phytotoxic products (e.g. acetic acid and ethanol are produced when organic matter is fermented in soil) and through changes to a variety of plant/microbial/animal interactions (e.g. anaerobic conditions favour the development of many root pathogens):

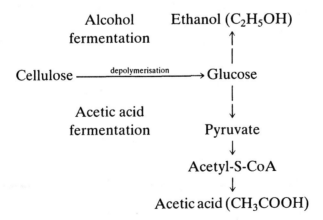

The phytotoxic effects of acetic acid, produced by organic matter fermentation under anaerobic conditions, are strongly pH dependent. Aliphatic acids such as acetic acid are most toxic at low pH because they are then lipophilic and therefore soluble in the lipid components of plant root membranes (Lynch, 1980). This is due to the fact that acetic acid (and similarly produced butyric acid) are weak acids and dissociate as follows:

$$CH_3COOH \rightleftharpoons CH_3COO^- + H^+$$

The pK_a of acetic acid is 4.75. This means that although nearly 90% of the acid is undissociated at pH 4, only 5% is undissociated at pH 6. One would therefore expect any phytotoxic effects of acetic acid to be minimal in most arable soils where pH is maintained close to neutrality.

Mineralisation of soil organic matter is generally retarded under anaerobic conditions, and considerable accumulation of undecomposed and partially decomposed plant residues can occur in the soil. Decomposition of a number of organochlorine xenobiotics, such as the pesticides DDT and Lindane, which are associated with pollution problems in some soils, is accelerated under anaerobic soil conditions (Skinner, 1975).

Soil organic matter and trace element availability

Although plant growth is most often limited by the availability of a major soil nutrient such as nitrogen, phosphorus or potassium, it is not uncommon to find growth limited by availability of minor (magnesium, sulphur, calcium) and even trace or micronutrients such as copper, manganese, zinc, boron, molybdenum and iron (iron is generally found in plant tissues at somewhat higher concentrations than the other trace nutrient elements).

The availability of trace elements to plants, which is usually estimated in the laboratory by some form of empirically selected chemical extraction (see Chapter 5), may be controlled to a considerable degree by interaction with soil organic matter.

Release of trace elements into soil solution

Trace elements are released into soil solution directly through the weathering of minerals and the decomposition of organic matter, as well as from ion exchange processes. Boron, for example, is slowly released from borosilicate minerals such as tourmaline, while manganese is released from secondary oxides such as pyrolusite (MnO_2) as well as from primary minerals. The decomposition of organic residues also releases the trace elements, although this aspect of cycling is generally secondary to release from minerals.

When released into the soil solution, the trace elements can occur in a variety of forms, depending on the solution chemistry. Copper, for example, can be found in soil solution as Cu^{2+}, $CuOH^+$ and $CuCO_3^0$, the latter species becoming more important as the solution pH increases and as respiration increasingly contributes carbon dioxide to the soil atmosphere respectively. For zinc, Zn^{2+} predominates below pH 7.7, with $ZnOH^+$ being the main species at higher pH. Equilibria of trace elements in solution will be discussed in Chapter 4.

Regardless of the dominant species present, trace elements are almost always found only in very low concentrations in the soil solution – often a few parts per billion. The main exception to this is at extremes of soil pH since, as seen earlier, the solubility of the trace elements is strongly pH dependent. The solubility of zinc, for example, can decrease 100 fold with each unit increase in soil pH (Tisdale, Nelson & Beaton, 1985), although solubility will also depend on the composition of the soil water and on the nature of the soil's cation exchange sites. It is because the concentrations of trace elements in solution are so low that contribution from decomposing organic matter can be significant, and therefore organic residue incorporation may alleviate trace element deficiencies. However, organic matter also serves as a major sink for trace nutrient elements.

The association of trace elements with soil organic matter may result in either increased or decreased availability. Reduced availability results from complexation with high molecular mass organics such as humic acids and lignin and complexation with initially soluble organics which then form insoluble precipitates. Increased availability can result from solubilisation and mobilisation by short chain organic acids, amino acids, and bases. Some of these associations between organic matter and trace elements are discussed in the following sections.

Depletion of trace elements in soil solution by bonding to organic matter

Cationic micronutrient element solution concentrations are modified by adsorption to the many cation exchange sites of organic matter, where the nutrients are held electrostatically in response to Coulombic forces. The cation exchange capacity of soil humus can exceed 300 milliequivalents per hundred grams and it has been estimated that between 20 and 70% of the soil's cation exchange capacity typically results from organic matter (Stevenson, 1982). A considerable trace element pool is held on these organic sites and may be released by simple cation exchange.

Soil organic matter

Chelation by organic ligands

The water molecules surrounding a trace element ion may be replaced by other molecules or ions to form a coordination compound. The particular molecule or ion which combines with the trace element ion is known as a ligand and the resulting complex is a chelate if two or more of the coordinate positions around the trace ion are occupied by groups of a single donor ligand, forming an internal ring structure.

The tendency of a trace element ion to form a complex with an organic ligand (or any other type of ligand) is defined in terms of the formation or stability constant of the reaction:

$$M + L \rightleftharpoons ML$$

$$K = \frac{[ML]}{[M][L]}$$

where K is the formation or stability constant, [M] is the trace element concentration, [L] is the organic ligand concentration and [ML] is the concentration of metal–ligand complex.

The formation of a greater number of bonds between ligand and trace element generally increases the stability of the chelate, and hence the tenacity with which the trace element is held. The stability of the trace element chelate is also influenced by the nature of both metal ion and ligand, by the number of rings formed and by the pH of the soil solution. The different affinities of organic groups for trace element ions generally follow the order: carbonyl (C=O) < ether (–O–) < carboxylate (–COO–) < ring N (–N=) < azo (–N=N–) < amine (–NH$_2$) and enolate (dissociated phenolic -OH, O$^-$). Thus carbonyl groups have the lowest affinity and enolate groups the highest.

Of all the components of organic matter involved in trace element chelation in the soil, the most important quantitatively is soil humus (Linehan, 1985), largely because of its carboxyl, acid hydroxyl and nitrogen groups. Much of the evidence for the nature of bonding between humic components (humic acid, fulvic acid and humin) comes from electron spin and nuclear magnetic resonance spectroscopy (e.g. McBride, 1978).

In addition to humic components, exudates from plant roots are also known to be involved in trace element binding. The exudates consist, in particular, of amino acids and organic acids which both have considerable potential as trace element chelating agents (Martell, 1975). The rhizosphere, therefore, represents a very important zone of the soil in terms of trace element chelation. Organic chelation can enhance movement of trace elements to the root and may be a key feature of the strategy of trace element acquisition by plants in the rhizosphere (Lindsay, Hodgson & Norvell, 1966).

As well as organic matter affecting trace element availability through complexation and chelation processes, the processing of organic matter and other activities carried out by soil microorganisms can alter the redox potential and pH of the soil. These changes may also affect availability of trace elements by altering the solution concentration and modifying the distribution of species present for a given element.

Other roles of soil organic matter

Before leaving the subject of soil organic matter, it is important to mention briefly its role in improving the water holding capacity of soils, especially when the soil mineral matter is coarse textured. Both incorporation and surface mulching may substantially reduce water loss from soil. Finally, the most conspicuous effect of organic matter is to make a soil darker in colour. With practice it is sometimes possible to make a reasonable estimate of soil organic matter content from its appearance. Changes in colour also influence the thermal absorption and radiation characteristics of a soil. A darker soil will cool faster at night than a lighter soil at the same temperature. It will also heat up more rapidly during the day.

4

Soil chemical reactions

Having considered at some length in Chapters 2 and 3 the nature and origins of both the inorganic and organic components of soil, we are now in a position to consider how these components interact with water to regulate the chemical composition of the soil solution. These chemical and physicochemical interactions govern the chemical composition of the plant root environment, and hence short-term nutrient availability to plants. They also regulate drainage water chemistry, and are therefore clearly of great importance. It is appropriate to begin by looking more closely at the cation exchange behaviour of soils.

Cation-exchange properties of soil clays

The existence of cation-exchange properties of smectites and vermiculite has already been mentioned in Chapter 2. So far, however, we have only considered cation exchange from negatively charged sites arising as a consequence of isomorphous substitution by cations of similar size but with lower charge, occurring in interlayer spacings. Generally the cations in the interlayer spacings are only labile when the spacing is *ca* 1.4 nm or more. In addition to negative charge arising as a result of isomorphous substitution, it also occurs at the edges of crystals where valences would otherwise be incompletely satisfied. The individual clay particles in a moist soil will thus be surrounded by a layer of hydrated, positively charged cations. This, in turn, creates a localised zone of positive charge, which will attract anions from the soil solution, creating an electrical double layer. The solute concentration in the immediate vicinity of a clay crystal therefore exceeds that in the bulk solution. We shall return to this point later, when discussing the measurement of soil pH and the structural stability of soils in the field when subjected to diverse management.

The cation-exchange capacity of a soil is also of vital importance in assessing the amount of acidity stored in it, or the amount of lime required to change its pH (see Chapter 5). It is also invaluable when assessing the fate of fertilisers applied to a soil and, as we shall also see in Chapter 5, a knowledge of the clay mineral composition also then becomes important.

Inner-sphere and outer-sphere surface complexes

When considering the nature of interlayer bonding in the case of micas and vermiculite, it was stated that the layers are held together strongly by potassium cations. When some of the silicon atoms in the tetrahedral sheet are replaced by aluminium atoms with less positive charge, the surplus negative charge created is delocalised over three oxygen atoms of one silicon tetrahedron (Sposito, 1984). Much stronger complexes with cations and dipolar molecules are formed between layers. Such a complex is known as an inner-sphere complex. It will be seen in Chapter 5 that this mechanism of potassium fixation is very important when considering the fate of added potassium fertilisers.

When isomorphous substitution of Mg^{2+} or Fe^{2+} for Al^{3+} occurs in an octahedral sheet, the surplus negative charge created is delocalised over ten outer oxygen atoms of four silica tetrahedra. Conditions are thus less favourable for a strong complex of the inner-sphere type described above. The result is the formation of a weaker, outer-sphere complex with hydrated cations which are therefore relatively more mobile, and readily exchangeable. Further details of the nature of bonding at mineral surfaces are outside the scope of this work, but may be found in the useful text on the surface chemistry of soils by Sposito (1984).

Measurement of exchangeable cations and exchange capacity

Although this book is not concerned primarily with soil chemical analysis, it is useful at this point to consider briefly some of the methods used to quantify the cation exchange properties of soils. An understanding of the techniques used should help to clarify the physico-chemical mechanisms involved and the importance of exchangeable cations and cation capacity (CEC) in soil chemistry. Some of the cation exchange sites in soils are associated with soil organic matter rather than clay minerals. This aspect was dealt with in Chapter 3. Because the dissociation of phenolic (OH) and carboxylic acid (COOH) groups on soil organic matter is strongly pH dependent, it follows that CEC also varies with pH in many soils.

If a sample of soil is mixed with a solution containing a high concentration of an appropriate cation such as ammonium, the ammonium will quite rapidly displace substantial amounts of the other cations adsorbed onto cation exchange sites. The greater the excess of ammonium ion concentration over displaced cation concentration in solution, the greater the efficiency of displacement. Thus the amount of each different cationic species remaining on exchange sites depends upon the extent of its initial occupancy of the exchange sites and the concentration of ammonium ions in the initial solution. If the soil sample is then filtered or centrifuged and reextracted with a second portion of ammonium solution, the concentrations of displaced cations in the equilibrating solution will be lower than in the first extraction, and therefore the overall efficiency of the extraction is improved. In practice three extractions are generally sufficient to effect virtually complete extraction of exchangeable cations. The three extracts are combined, diluted to a known volume, and the resulting solution analysed by atomic absorption and/or emission spectrometry (Ure, 1991).

The cation-exchange capacity of the soil may then be measured by washing out the non-adsorbed ammonium ions and displacing the adsorbed ammonium ions with a suitable solution such as weakly acidified sodium chloride. Measurement of the displaced ammonium gives a measure of the CEC.

Multiple shake-and-centrifuge techniques are rather time-consuming and labour intensive. In many laboratories, therefore, a continuous leaching technique is preferred. In the authors' Department, sets of 12 of the units shown in Fig. 4.1 are used. Samples are leached first with ammonium acetate, then with aqueous ethanol or propan–2-ol, and finally with acidified sodium chloride solution. The chicken-feed system (so-called because it allows unattended refilling of water trays for chickens, etc.) allows completely unattended operation.

The procedure described above is not suitable for all soil types. Soils containing significant amounts of gypsum or calcium carbonate, or, for arid areas, soluble salts, require different procedures, as do soils with significant vermiculite contents. Although ammonium ions displace exchangeable base cations from vermiculite, they are not then themselves displaced by the acidified sodium chloride. Details of procedures for such problem soils may be found in a comprehensive text on soil analysis (Rhoades, 1982a).

Some theoretical aspects of cation exchange

Although strictly speaking it is still not possible to describe quantitatively cation-exchange reactions at a fundamental ionic process level (Sposito, 1984), there are useful experimentally-derived descriptive models that are valuable for predictive purposes (Russell, 1987).

The Ratio Law

For a soil equilibrating with an aqueous solution to yield a system containing only two cationic species of similar valence at

Fig. 4.1 Leaching apparatus for the determination of exchangeable cations and cation-exchange capacity.

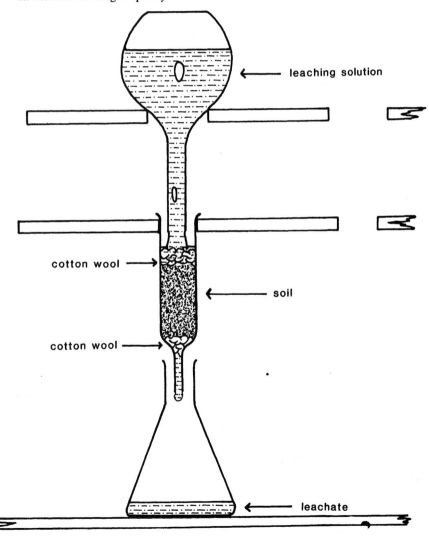

significant concentrations, the relative amounts adsorbed depend upon their relative amounts (to be precise, upon the ratio of their activities) in the equilibrating solution. If the ions have a different valence, for example the M_1^{n+}/M_2^{m+} system, the adsorption ratio depends upon the ratio of the mth and nth roots of the activities of M_1^{m+} and M_2^{n+} respectively. This law is known as the Ratio Law. It has been shown to be valid for a range of ion pairs and for the ratio of potassium activity to the square root of the sum of the activities of calcium and magnesium (Russell, 1987).

The ratio in which ions are adsorbed from a mixture of cations in the equilibrating solution depends also upon the nature of the cationic species and of the adsorbing substrate. Thus, for M_1^{m+} and M_2^{n+}:

$$\sqrt[m]{\{M_1\}}/\sqrt[n]{\{M_2\}} = k\,[M_{1ex}]/[M_{2ex}]$$

where $\{M1\}$ and $\{M2\}$ refer to the activities of species M_1^{m+} and M_2^{n+} in solution respectively and $[M_{1ex}]$ and $[M_{2ex}]$ to the concentrations of M_1^{m+} and M_2^{n+} on cation exchange sites, respectively. Thus k is a measure of the relative adsorbing power of a pair of cations for a given soil. This equation is known as Gapon's Equation. A number of improved equations have been suggested to describe cation exchange behaviour, but that given above is sufficient here. More precise models of cation exchange are academically interesting, but it must be remembered that, in practice, the situation is complicated by competing effects of microbial and plant uptake and inputs from geochemical weathering and other sources. This will become clearer when we consider the factors regulating soil acidity in the next section.

Particle size distribution in soils

The major contributions to the CEC of soils come, as we have seen, from exchange sites in the interlayer spacing of certain crystalline clay minerals, the surplus charge at the surfaces of edges of crystals and the soil organic matter. The charge at the surface of particles of any specified composition obviously will increase with the surface area exposed, which, in turn, increases with a decrease in the mean particle diameter. Thus, other factors being equal, the smaller the particles, the greater the surface area and the greater the CEC. The particle size distribution of the mineral grains in soils is therefore of great relevance to its cation exchange properties. It also regulates to a large degree the ease with which a soil may be cultivated and its drainage characteristics, aspects considered further in Chapter 5. It is a simple matter to demonstrate the relationship between surface area and particle size if we assume for simplicity that the particles are spherical. The ratio, R, of surface area to volume for a single particle of radius r is given by:

$$R = 4\pi r^2 / (4/3)\pi r^3 = 3/r$$

When r is 1 mm, R equals 3 mm^{-1}, and when r is 0.001 mm, or 1 μm, R equals 3000 mm^{-1}. At a given density, mass is directly proportional to volume. If samples with the same mass of two soils with the same particle density are compared, then the total particle volume is the same in both samples. Thus R values may be compared to see how relative surface area varies with relative r values. Figure 4.2 is a graph of R against r, plotted on logarithmic graph paper. It shows that, to a first

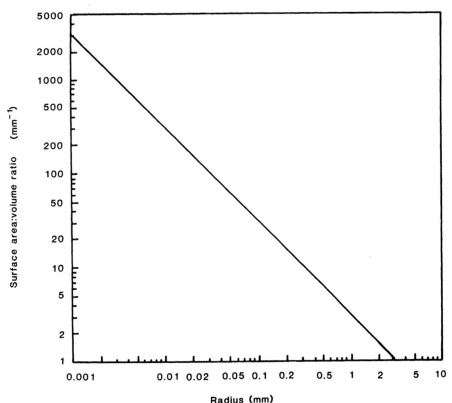

Fig. 4.2 Relationship between surface area-to-volume ratio and particle radius for a solid comprised of identical spheres.

approximation, if the particles of a mineral are a thousand times smaller in diameter in one sample than in another, the finer particles might be expected to have a thousand times greater surface charge per unit mass.

Figure 4.3 shows three major systems used to classify particle size ranges (after White (1987)). The proportion of the total mass of soil falling in each of the size ranges is usually measured in the laboratory by a sedimentation technique of some sort, larger particles settling out more rapidly than smaller ones. Calculation of results is invariably

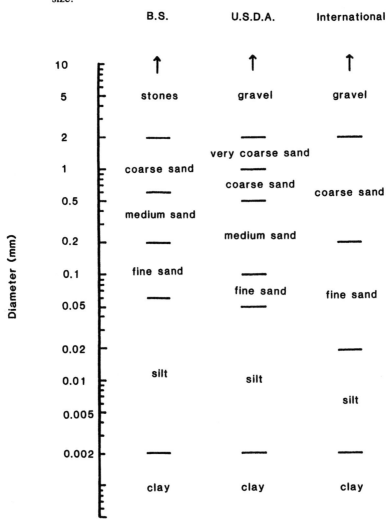

Fig. 4.3 Three major systems for classifying soil particles according to their size.

based on the assumption that the particles may be considered as free-falling spheres in a viscous fluid, so the diameters in Fig. 4.3 are in practice equivalent spherical diameters.

It will be seen from Fig. 4.3 that the smallest size fraction is known as the clay fraction. It contains all the particles less than 2 μm in diameter in all three classification systems. The term 'clay fraction' should not be confused with the term 'clay minerals'. The clay fraction may, and often does, contain clay-sized particles which are not clay minerals. Similarly clay minerals may occur in particles which are too large to be included in the clay fraction. Some authors use the term 'phyllosilicates' to describe the layer silicates in an attempt to avoid confusion (see, for example, Greenland & Hayes, (1978)). Since however the term 'clay minerals' is so widely used throughout the literature of the subject, the avoidance is, at best, short term.

Soil pH

From the preceding discussion of cation-exchange properties of soil, it should be clear that some of the cation-exchange sites of soils may be occupied by hydrogen ions and that, when the soil equilibrates with an aqueous solution, some of these hydrogen ions will pass to the solution phase. The concentration of hydrogen ions in solution is expressed, because the range of concentrations encountered covers more than 14 orders of magnitude, (1 M–10^{-14} M), on a logarithmic scale known as the pH scale. The pH of a solution is defined as the negative logarithm to the base 10 of the hydrogen ion activity, i.e.:

$$\mathrm{pH} = -\log\{\mathrm{H}^+\}$$

where $\{\mathrm{H}^+\}$ denotes the activity of hydrogen ions. In practice, for very dilute solutions, $\{\mathrm{H}^+\}$ and the hydrogen ion concentration are approximately the same. Thus a 10^{-7} M solution of H^+ ions has a pH value of 7. This corresponds to the pH value of pure deionized water, which has a dissociation constant of 10^{-14}. Thus:

$$K_w = \{\mathrm{H}^+\}\{\mathrm{OH}^-\} = 10^{-14}$$

If ionization of water is the only source of H^+ and OH^-, it follows that $\{\mathrm{H}^+\}$ and $\{\mathrm{OH}^-\}$ must be equal to maintain electroneutrality, thus:

$$\{\mathrm{H}^+\} = \{\mathrm{OH}^-\} = 10^{-7}$$

A pH value of 7 is therefore generally regarded as neutral. In soil chemistry it is rare to encounter natural solutions in which the hydrogen ion concentration falls outside the range 10^{-2} M–10^{-10} M (pH 2 to pH 10).

Measurement of soil pH

Since pH is, by definition, a measure of the activity of hydrogen ions in a solution, it is clearly not possible to measure the pH of a dry soil. As soon as we accept that it is necessary to add water to soil prior to pH measurement, we are confronted with a number of difficulties, however. The first problem is deciding just how much water to add. Soils differ tremendously in their capacity to absorb water. We have already seen that some clay minerals can absorb water into their interlayer spaces, with consequential swelling. In addition, when the clay-sized fraction of a soil is substantial, the soil tends to form aggregates which will absorb substantial quantities of water into small pores. It may therefore be necessary to add water to give a higher soil:water ratio for a dry soil with a high clay content than for a dry sandy soil. Soil organic matter too can absorb substantial quantities of water. Thus some soil chemists prefer simply to prepare a thick paste, rather than to work at a fixed water:soil ratio. This provides conditions which might be regarded as more realistic, but at the expense of reducing precision.

A further complication arises because of the electrical double layer which was mentioned briefly in the section on cation-exchange properties of soil clays. Thus surplus negative charge on mineral particle surfaces is balanced by adsorbed exchangeable cations. These, in turn, create a zone of weak positive charge which attracts anions. The net effect is that the concentration of ions in the bulk solution may differ from that in the immediate vicinity of the particles. The situation is further complicated by the nature of ion exchange equilibria themselves. The greater the concentration of any cation other than H^+ in solution, the more it will be adsorbed onto the cation-exchange sites, and the more H^+ ions will be displaced into solution. The measured pH value therefore depends to some extent upon the soluble salt content of a soil, regardless of whether the salts are derived from geochemical weathering in the soil, from fertiliser, or from inputs, perhaps of maritime origins, in rainwater or atmospheric aerosols. One consequence of this is that, as an air-dried acid soil equilibrates with distilled water over about 60 minutes, the measured pH value tends to fall as the concentrations of soluble salts increase slightly and more H^+ ions are displaced from exchange sites. Equilibrium is attained rather more rapidly if field-moist soil is analysed. The situation may be exacerbated by the fact that dissolved carbon dioxide originating from microbial respiration produces carbonic acid, which contributes to the geochemical weathering-derived soluble salt (as HCO_3^- salts) content. As a result of these complications, it is common practice when measuring soil pH to add a dilute calcium chloride solution, rather than distilled water, to the dry soil to reduce variability.

Soil pH is undoubtedly the most widely measured soil parameter, in spite of the problems outlined above. As will become clear later, not only can pH influence plant growth directly, but it also regulates the availability to plants of all the major and trace nutrient elements in soil. The mechanism of this control may be direct physicochemical adsorption or precipitation, or it may be *via* effects upon microbial activity. In very acid soils, aluminium may be mobilized in toxic quantities, causing severe stunting of root elongation, especially in sensitive plant species (see also Chapter 5).

The transient nature of soil pH

We have already seen that soil pH may change in the short term. However long-term changes, both from year to year and century to century, also occur. Figure 4.4 illustrates the factors which regulate the fraction of the CEC of a soil which is occupied by the base cations, Ca^{2+}, Mg^{2+}, Na^+ and K^+, a parameter known as the base saturation. At any stage in time, the pH of a soil thus depends upon the extent to which inputs of base cations from the atmosphere, from geochemical weathering, from plant litter, from fertilisers and manures and in water flowing into the soil from elsewhere can keep pace with outputs in drainage water, plant uptake, and crop and animal removal. Soils in regions with high rainfall, where water drains regularly through and from the soil, are thus more susceptible to long-term acidification than soils in drier areas. Indeed, in regions with very dry climates, not only may leaching losses be negligible, but salts may accumulate from rising water, resulting in salinity and sustained high soil pH values. On a seasonal timescale, high levels of base cation uptake by an annual crop may lead to temporary acidification over the growing season. Removal of cations in a harvested product can result in permanent acidification. Cations in crop residues, e.g. roots, straw, are returned to the soil *via* microbial degradation of the plant litter. It should never be forgotten that soil is a dynamic system, especially when samples are being taken for analysis.

68   Soil chemical reactions

Depending on the reason for the measurement being performed, the sample must be representative not only with respect to the area of interest, but also to depth and time.

Soil pH and nutrient availability

The essential plant nutrient elements apart from carbon, hydrogen and oxygen are primarily supplied from the soil. These three, which usually make up more than 90% of the mass of fresh plant tissue, differ in that they come from atmospheric carbon dioxide or from water. The soil-derived essential elements, and their important forms in soils, are listed in Table 4.1 (Jones, 1982). 'Essential' in the present context implies that the plant is incapable of completing every stage of its growth cycle in the absence of the element concerned. In addition to the elements listed, some plants show reduced yields if the available supply of sodium is low, most notably sugar beet, mangolds and some brassicas (Cooke, 1967; Russell, 1987). There are strong links between pH of soil and the availability of each of the nutrient ions included in Table 4.1. Indeed, the beneficial effects of liming to raise soil pH may be as much, or even more, due to effects on nutrient element availability as to direct effects upon plant growth. This section summarises briefly the effects of pH on nutrient availability.

Fig. 4.4 Sketch showing the major components of the biogeochemical cycles of base cation elements.

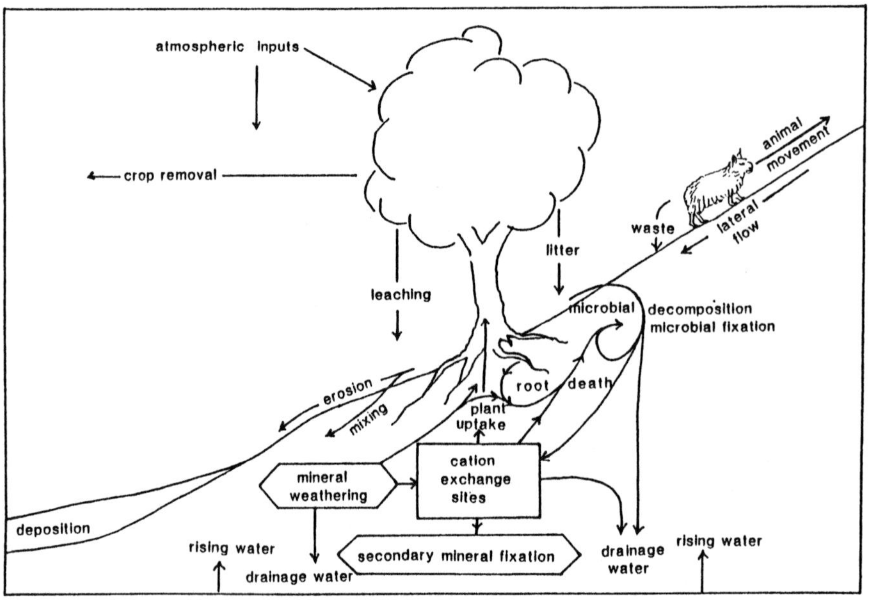

The base cations

It should be apparent from the earlier discussion of the dynamic nature of soil pH that acid soils are associated with low base saturation. Since the base cations on cation-exchange sites are, to a large degree, available to growing plants, it follows that the availability of calcium, magnesium, potassium and sodium is invariably lower in acid soils than in near-neutral or alkaline soils. That the available base cations tend to come virtually exclusively from the exchange sites in non-calcareous soils may be demonstrated by isotopic labelling techniques. For example, if a soil is equilibrated with a solution containing radioactive ^{45}Ca, some of the radioactive isotope exchanges with calcium on cation-exchange sites. If the soil is then washed, and used to grow plants, it may be shown that the ^{45}Ca specific activity is similar in the exchangeable soil fraction and in the plant (Russell, 1987).

Table 4.1. *Essential elements and their forms in soils*

Element	Cationic forms	Anionic forms	Other forms[a]
Nitrogen	NH_4^+	NO_3^-, NO_2^-	Organic-N
Phosphorus		HPO_4^{2-}, $H_2PO_4^-$ polyphosphates	Organic-P
Sulphur		SO_4^{2-}, S^{2-}	Organic-S
Potassium	K^+		
Calcium	Ca^{2+}		
Magnesium	Mg^{2+}		
Iron	Fe^{2+}, Fe^{3+}, $Fe(OH)^{2+}$, $Fe(OH)_2^+$		Organically complexed
Manganese	Mn^{2+}		Organically complexed
Copper	Cu^{2+}		Organically complexed
Zinc	Zn^{2+}		Organically complexed
Molybdenum		MoO_4^-	
Boron		$B(OH)_4^-$	H_3BO_3
Chlorine		Cl^-	[b]
Cobalt	Co^{2+}		
Selenium		SeO_4^{2-}, SeO_3^{2-}, Se^{2-}	Organic-Se

[a] Gaseous forms of nitrogen, sulphur and selenium in the soil atmosphere, and oxygen, hydrogen and carbon are not included.
[b] At high concentrations, some anionic metal chloride complexes may become significant.

In strongly alkaline soils, those with pH values above 8.5, sodium is invariably the dominant cation on the exchange sites, because calcium is precipitated as the carbonate before such high pH values are reached, i.e.:

$$Ca^{2+} + CO_2 (g) + H_2O \rightleftharpoons CaCO_3 + 2H^+ \qquad (4.1)$$

The calcium therefore has a buffering effect upon pH. Sodium has no such effect, and higher soil pH values may be obtained when sodium is the dominant exchangeable cation. It follows that soils with a pH value of 9 or above are often somewhat deficient in calcium and, for a similar reason, magnesium.

Copper, zinc and cobalt

These three elements are generally considered together because they are trace metal elements that exhibit similar trends in availability with changing soil pH (see, for example, Landon (1991)). However, the precise mechanisms causing the trends are not identical for all three. Cobalt is not generally regarded as essential for plant growth in the strictest sense, but is a very important element in animal nutrition, so it is worth including here.

In old, very acid soils (below pH 4.5), the availability of all three elements tends to be low. This is partly because the elements tend to be more soluble in acidic mineral soils, and may therefore be flushed out of the soil systems in drainage water. However, the situation is complicated by the fact that the activities of bacteria and actinomycetes are suppressed as soil pH declines. Litter decomposition becomes increasingly dependent upon the activity of fungi, and the decomposition rate slows down. This leads to organic matter accumulation in the soil. All three elements are quite strongly complexed by organic matter, copper in particular tending to form stable organic chelate complexes (see Chapter 3). This tends to reduce availability to plants in the short term, but may also have a long-term effect in thick, highly organic soils such as peats. If the plants are deficient in copper, for example, then the next generation of plant litter will be naturally low in copper, and this is the parent material of the newly forming peat. Thus, over hundreds of years, the total copper, and hence the available copper, tend to decline as the peat thickens. Trace element fertilisation may therefore be essential when peat-based composts are used in horticulture.

As soil pH increases towards pH 7, the availability of cobalt, copper and zinc starts to decline sharply. By examination of the zinc or copper concentrations in soil solution in equilibrium with soil at various pH values, Lindsay (1979) was able to calculate reaction constants for the soil used for the following reactions:

$$\text{soil-Zn} + 2H^+ \rightleftharpoons Zn^{2+} \qquad \log K^0 = 5.8 \qquad (4.2)$$

$$\text{soil-Cu} + 2H^+ \rightleftharpoons Cu^{2+} \qquad \log K^0 = 2.8 \qquad (4.3)$$

Such empirical constants, which will vary slightly from soil to soil, may be used to relate copper or zinc activities in soil solution to soil solution pH value:

$$\log \{Zn^{2+}\} = 5.8 - 2pH \qquad (4.4)$$

$$\log \{Cu^{2+}\} = 2.8 - 2pH \qquad (4.5)$$

It should be stressed that equations (4.4) and (4.5) are empirical rather than theoretical equations, but they show how the activities of copper and zinc decrease with increasing pH. The results for copper are shown in Fig. 4.5, together with the log $\{Cu^{2+}\}$ *versus* pH plots for some other common copper minerals, based upon data from Lindsay (1979). Results such as those shown in Fig. 4.5 are valuable insofar as they demonstrate that no simple single copper mineral can, on its own, be limiting the solubility of copper for the soil or soils under investigation, but that possibly a mixed amorphous iron–copper oxide could be responsible. This approach is developed further in Lindsay's excellent text, *Chemical Equilibria in Soils* (Lindsay, 1979).

In the case of zinc, the mixed oxide franklinite ($ZnFe_2O_4$) yields a value for $\{Zn^{2+}\}$ even lower than that found in soil solutions. Work in the authors' laboratories strongly suggests that phosphate is also involved in limiting zinc availability, and increasingly so with increasing soil pH, together with precipitating iron and aluminium hydroxides. A possible structure is shown below:

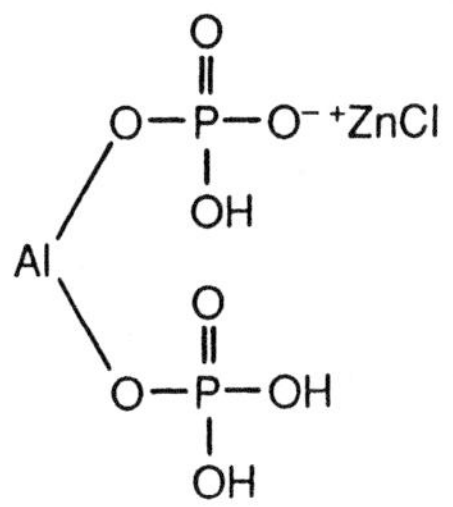

$$\text{Al} \begin{cases} O-\overset{\overset{\displaystyle O}{\|}}{\underset{\underset{\displaystyle OH}{|}}{P}}-O^-\ {}^+ZnCl \\ \\ O-\overset{\overset{\displaystyle O}{\|}}{\underset{\underset{\displaystyle OH}{|}}{P}}-OH \end{cases}$$

Soil chemical reactions

Figure 4.6 shows how zinc concentration in simple aqueous solutions varies with pH in the absence and presence of iron and aluminium (Jahiruddin, Chambers, Livesey & Cresser, 1986). These results clearly show that individual elements should not be considered in isolation when the effects of soil pH are being considered.

Iron and manganese

The soil solution chemistry of iron and manganese is complicated by the fact that both elements exist in more than one oxidation state in soils, iron in the Fe(II) and Fe(III) states and manganese in the Mn(II), Mn(III) and Mn(IV) states. Thus, while both elements tend to

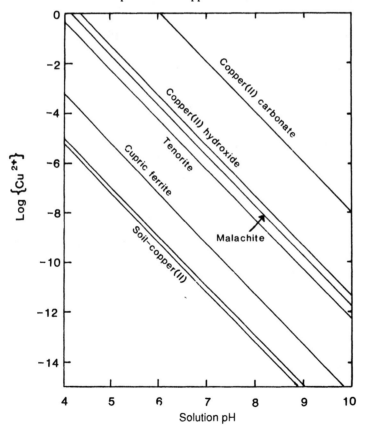

Fig. 4.5 The relationships between copper concentration in equilibrating solution and solution pH. The line closest to that for soil–copper(II) is for mixed amorphous iron–copper oxide.

Soil pH and nutrient availability

become less available as soil pH increases, regardless of oxidation state, the lower oxidation states tend to produce higher soil solution concentrations than the higher states. The lower states are found in anaerobic zones of soils where reducing conditions prevail. This aspect is considered further later, and was also discussed in Chapter 3. Manganese deficiency in plants is not uncommon in England when the soil pH exceeds 6.5. Further north, in Scotland, deficiency symptoms may be found even at pH 6.2. It appears that manganese deficiency is more probable in soils with a higher organic matter content and, generally, higher rainfall and lower temperatures in the north favour lower microbial activity and enhanced organic matter accumulation.

In acid soils, the total contents of iron and manganese are generally sufficient for them to be readily available, although their mobility in such soils may be sufficient for deficiencies to occur in very old soils, especially if the parent rock is of very low base status.

Molybdenum

Unlike the elements which we have considered so far in this section, the important form of molybdenum in soil with respect to plant uptake is anionic rather than cationic, and it is generally accepted that the molybdate ions are largely adsorbed onto the surface of hydrous iron oxides in the soil. The extent of adsorption decreases with increas-

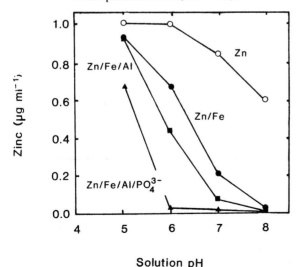

Fig. 4.6 Diagram showing the effect of pH on the solubility of Zn alone, and in the presence of Fe, Fe + Al, and Fe + Al + PO_4^{3-}.

ing pH over the range 4.45–7.75 (Follett, Murphy & Donahue, 1981; Gupta & Lipsett, 1981). This introduces a new concept, namely that of anion exchange. In the case of molybdate, the anion effectively displaces a hydroxyl from the hydrated oxide. However, other anions, such as sulphate, will compete with the molybdate ions for adsorption sites. Indeed, sulphate additions may be used to regulate molybdenum availability to plants (Gupta & Lipsett, 1981). We shall return to the important topic of anion exchange later in this chapter and others. Molybdenum continues to increase in availability with increasing soil pH.

Phosphorus

Phosphorus, like molybdenum, exists in soils in anionic forms, which, again in common with molybdate, are strongly adsorbed under acid conditions by hydrous oxides of iron and aluminium. This process has already been mentioned briefly when the effect of pH on zinc availability was discussed. If the soil is sufficiently acidic for the soluble iron and aluminium to be significant, then these cations will precipitate out phosphate. The precipitate becomes increasingly less soluble with aging (Brady, 1990). Anion adsorption also occurs, but rather less strongly, at the surface of silicates, and this is the dominant adsorption process near or at neutral pH. As soil pH is increased from very acid values (*ca* pH 3) to near neutrality (*ca* pH 7), phosphorus availability therefore increases steadily. This is exacerbated by the increased microbial activity as the soil pH increases, which results in increased mineralisation (conversion from an organic to an inorganic form) of some of the organic phosphorus. However, above pH 7 the analogy to molybdate breaks down, because sparingly soluble calcium phosphate is precipitated in the presence of the high calcium concentrations usually present at circumneutral/moderately high soil pH. The reactions regulating phosphorus availability over the complete range of soil pH values are summarised in Fig. 4.7. As mentioned earlier, when the soil pH exceeds 8.5, sodium dominates the cation exchange complex, and precipitation of calcium phosphate is less likely.

Before leaving the topic of pH effects on phosphorus availability, it is worth briefly considering how the dissociation of phosphoric acid, H_3PO_4, changes with solution pH. The dissociation reactions, and the appropriate dissociation constant data (Lindsay, 1979), are as follows:

Soil pH and nutrient availability

$$H_3PO_4 \rightleftharpoons H^+ + H_2PO_4^- \qquad \log K = -2.15 \qquad (4.6)$$

$$H_2PO_4^- \rightleftharpoons H^+ + HPO_4^{2-} \qquad \log K = -7.20 \qquad (4.7)$$

$$HPO_4^{2-} \rightleftharpoons H^+ + PO_4^{3-} \qquad \log K = -12.35 \qquad (4.8)$$

It is possible to express each of the species present in terms of $H_2PO_4^-$. Rewriting equation (4.6) in the standard mass-action form familiar to most chemists gives:

$$\{H^+\}\{H_2PO_4^-\}/\{H_3PO_4\} = 10^{-2.15} \qquad (4.9)$$

or

$$\{H_3PO_4\} = \{H^+\}\{H_2PO_4^{--}\}/10^{-2.15} \qquad (4.10)$$

Fig. 4.7 Reactions regulating the solubility of phosphate over the pH range encountered in soils.

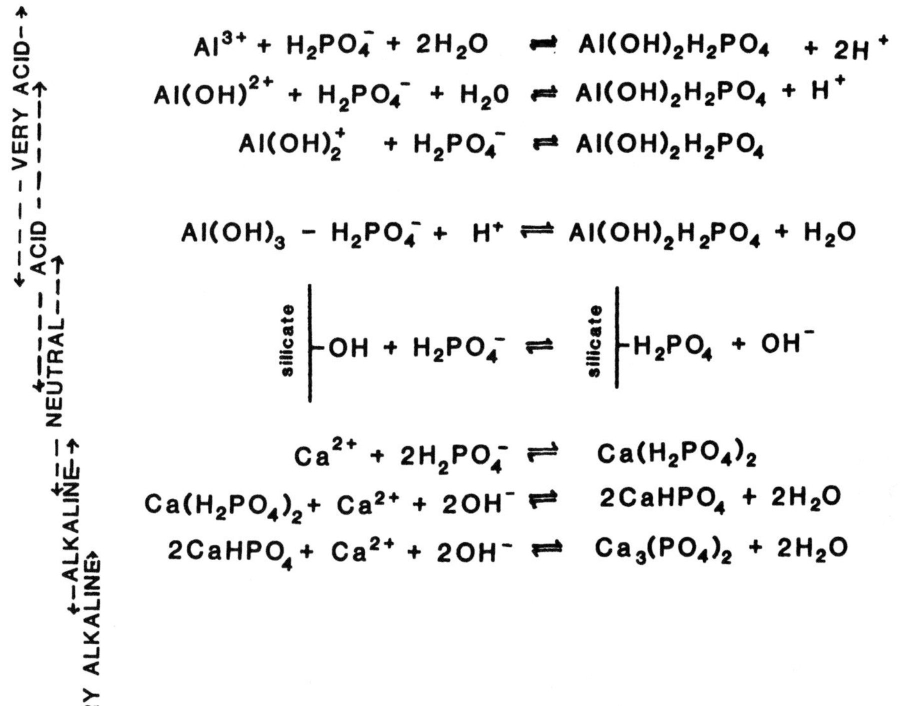

$$Al^{3+} + H_2PO_4^- + 2H_2O \rightleftharpoons Al(OH)_2H_2PO_4 + 2H^+$$

$$Al(OH)^{2+} + H_2PO_4^- + H_2O \rightleftharpoons Al(OH)_2H_2PO_4 + H^+$$

$$Al(OH)_2^+ + H_2PO_4^- \rightleftharpoons Al(OH)_2H_2PO_4$$

$$Al(OH)_3 - H_2PO_4^- + H^+ \rightleftharpoons Al(OH)_2H_2PO_4 + H_2O$$

$$\text{silicate}{-}OH + H_2PO_4^- \rightleftharpoons \text{silicate}{-}H_2PO_4 + OH^-$$

$$Ca^{2+} + 2H_2PO_4^- \rightleftharpoons Ca(H_2PO_4)_2$$

$$Ca(H_2PO_4)_2 + Ca^{2+} + 2OH^- \rightleftharpoons 2CaHPO_4 + 2H_2O$$

$$2CaHPO_4 + Ca^{2+} + 2OH^- \rightleftharpoons Ca_3(PO_4)_2 + 2H_2O$$

← VERY ACID — ACID — NEUTRAL — ALKALINE — VERY ALKALINE →

Similarly, from (4.7):

$$\{HPO_4^{2-}\} = 10^{-7.20}\{H_2PO_4^-\}/\{H^+\} \quad (4.11)$$

and from (4.8):

$$\{PO_4^{3-}\} = 10^{-12.35}\{HPO_4^{2-}\}/\{H^+\} \quad (4.12)$$

Combining (4.11) with (4.12) gives:

$$\{PO_4^{3-}\} = 10^{-7.20} \times 10^{-12.35}\{H_2PO_4^-\}/\{H^+\}^2 \quad (4.13)$$

The mole fraction of $H_2PO_4^-$, MF, is given by its concentration divided by the sum of the concentrations of all the phosphate species present, each of which may be expressed in terms of $H_2PO_4^-$. Thus:

$$MF = \frac{1}{10^{2.15}\{H^+\} + 1 + 10^{-7.20}/\{H^+\} + 10^{-19.55}/\{H^+\}^2} \quad (4.14)$$

from which it is possible to calculate the mole fraction of $H_2PO_4^-$ at any specified pH. The same approach may be employed to calculate the mole fractions of the other species present. The results are most conveniently expressed in graphical form. This has been done in Fig.

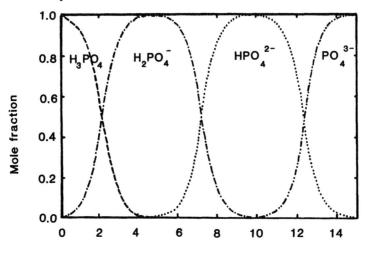

Fig. 4.8 Anionic forms of phosphorus in soil solution as a function of soil pH.

4.8, which shows that $H_2PO_4^-$ is the dominant form in solution for most soils, although HPO_4^{2-} becomes important in alkaline soils.

Boron

Like molybdenum and phosphorus, boron exists in soils in an anionic plant-available form. It is a major component of the mineral tourmaline, but this is not an important source of the element because it is such a stable mineral. The effect of pH upon boron availability is of particular importance because deficiency of the element may have a disastrous effect upon crop yield, quality and marketability (Shorrocks, 1974). Plants differ markedly in their sensitivity to boron deficiency (Follett, Murphy & Donahue, 1981). A further complicating factor is that the plant tolerance range for boron is very low, toxicity symptoms occurring at concentrations only slightly greater than those associated with deficiency symptoms. Toxicity symptoms associated with excess of the element were recognised before it was realised that it was essential (Thompson & Troeh, 1978).

Borates are generally quite soluble, but the anion is retained in association with soil organic matter, probably *via* formation of stable complexes with diols or polyphenols, and by adsorption onto hydrous oxides of iron and aluminium. The element is generally readily available in acid soils, but not in alkaline soils. Indeed, boron deficiency may often be one of the most serious consequences of overliming in agriculture. The authors have also seen boron deficiency induced in trees by overliming of forest soils in attempts to combat possible adverse effects of acid deposition. The precise mechanism by which the element is rendered unavailable at high pH requires further research.

Nitrogen

Although nitrogen occurs in soils in both cationic (ammonium, NH_4^+) and anionic (nitrate, NO_3^- and nitrite, NO_2^-) forms, the greater part by far occurs in organic forms, as discussed in Chapter 3. Of the 5% or so in the inorganic form, nitrite, which is phytotoxic, is generally negligible. Ammonium tends to be held on cation-exchange sites, and may be fixed by vermiculite, as we saw earlier. Nitrate is a highly mobile anion, and is not substantially adsorbed by soil mineral components at any pH. However, unless the soil contains a large pool of readily available nitrogen, nitrate tends to be taken up by the soil microbial biomass or by plants during the active growth period (see Chapter 3).

As a consequence, nitrogen availability tends to depend upon the rate at which organic nitrogen is converted to inorganic nitrogen, a process known as mineralisation. Mineralisation rate is temperature and pH-dependent, and bacterial oxidation of ammonium to nitrate especially is reduced under acidic conditions. The net result is that nitrogen also tends to be less available when the soil pH falls to below 5.5–6. Mineralisation rate also declines above pH 8.

Sulphur

In the soils of arid regions where leaching is generally negligible, sulphur tends to accumulate as the sulphate anion, SO_4^{2-}, which is the readily plant-available form of the element. Thus sulphur deficiency is rare in soils with a pH value above 8. In acid soils with a pH below ca 5.5–6, sulphate tends to be adsorbed onto the surface of hydrated iron and aluminium oxides, effectively exchanging for hydroxyls. In this respect its behaviour is analogous to that of phosphate although sulphate is less strongly adsorbed.

In the upper horizons (recognisable individual layers) of mineral soils, the greater part of the sulphur generally occurs in organic forms. Mineralisation of organic sulphur, like that of organic nitrogen, tends to be pH-dependent, the rate declining as the pH falls (Germida, Wainwright & Gupta, 1992). Therefore the availability to plants of both inorganic and organic sulphur falls with decreasing pH.

Other essential elements

We have discussed at some length over the preceding pages the influence of soil pH upon the plant availability of all of the elements which are universally accepted as essential. Chlorine and selenium are also probably essential in minute amounts (Bowen, 1979). Chloride is ubiquitous, the small amounts required invariably coming from aerosols of maritime spray origin even in inland areas. The great mobility of the chloride anion at all pH values is therefore of little consequence. Selenium has attracted relatively little attention except in regions with selenium-rich soil where grazing toxicity problems may be encountered (Jeffrey, 1987). However the element is now known to be of great interest in animal and human nutrition too. Our understanding of the behaviour of soil selenium is growing steadily, in part at least because of the development of a very sensitive method for the determination of the element, namely hydride generation atomic absorption spectrometry. Selenate might be expected to show similar trends with pH as sulphate.

Iodine and silicon are thought to be essential to some plants, and arsenic, bromine, chromium, fluorine, nickel, tin and vanadium may be essential, although further investigations are needed to establish irrefutably the requirement for these elements (Bowen, 1979). They will not be considered further here.

Saline and sodic soils

Although excessive salinity (accumulation of soluble salts) and alkalinity are, in fact, quite separate problems insofar as either on its own may adversely affect the growth of plants, they tend to occur under similar climatic conditions. Moreover, bad management of saline soils may eventually lead to soils which are both saline and excessively alkaline (pH 8.5–10). It is convenient therefore to consider both problems together at this point.

Saline soils

It was mentioned earlier in this chapter, in the section on base cations, that as long as calcium dominates the cation exchange complex rather than sodium, the soil pH is unlikely to rise above *ca* 8.5. It was stated that this is so because of the buffering effect of the precipitation of calcium carbonate (equation (4.1)). We now need to consider this aspect in more detail.

The effect of pH upon the dissolution of calcite ($CaCO_3$) may be calculated from the appropriate equilibrium constant (Lindsay, 1979):

$$CaCO_3 + 2H^+ \rightleftharpoons Ca^{2+} + CO_2(g) + H_2O \quad \log K = 9.74 \tag{4.15}$$

Thus:

$$\{Ca^{2+}\}\{CO_2\}\{H_2O\}/\{CaCO_3\}\{H^+\}^2 = 10^{-9.74} \tag{4.16}$$

Since both $\{H_2O\}$ and, when solid calcite is present, $\{CaCO_3\}$ equal unity:

$$\{Ca^{2+}\}\{CO_2\}/\{H^+\}^2 = 10^{-9.74} \tag{4.17}$$

Taking logarithms on both sides, substituting pH for $-\log\{H^+\}$, and rearranging:

$$\log \{Ca^{2+}\} = 9.74 - 2pH - \log \{CO_2\} \tag{4.18}$$

80   Soil chemical reactions

Figure 4.9 shows how log $\{Ca^{2+}\}$ varies with pH when the soil atmosphere carbon dioxide partial pressure is 0.01 or 0.001 (i.e. 1% or 0.1% by volume in the soil atmosphere). At pH 8, for example, and 1% carbon dioxide, log$\{Ca^{2+}\}$ is equal to -4.26, or $\{Ca^{2+}\}$ is $10^{-4.26}$ M. Thus, when the calcium on cation exchange sites is sufficient to give a $\{Ca^{2+}\}$ value of a few mg l^{-1} in the soil solution, calcium carbonate starts to precipitate. The equilibrium represented by equation (4.15) starts to move from right to left, effectively buffering the soil pH. Another calcium carbonate mineral, aragonite, is only very slightly more soluble than calcite, and in practice both calcite and aragonite may be found together in soils (Lindsay, 1979).

Sodium carbonate is orders of magnitude more soluble than calcite. In arid areas where the sodium input from the atmosphere or from irrigation water is too high, there is no mechanism for removal of the

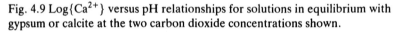

Fig. 4.9 Log$\{Ca^{2+}\}$ versus pH relationships for solutions in equilibrium with gypsum or calcite at the two carbon dioxide concentrations shown.

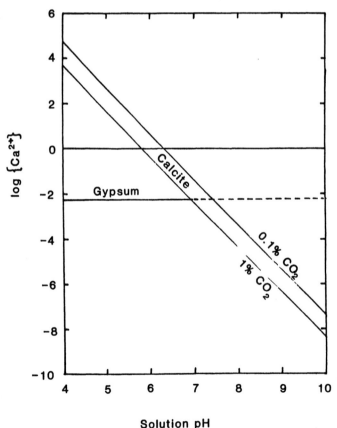

high concentration of sodium salts in the soil solution analogous to that described above for calcium. Thus sufficient soluble salt may accumulate to damage plant growth. The same is true for potassium. The associated anions will be carbonate, bicarbonate, chloride, sulphate and small amounts of nitrate and borate.

The soluble salt content of a soil is normally described in terms of the electrical conductivity when it is converted to a saturated paste at 25 °C. The electrical resistance of soil solutions decreases with increasing dissolved salt concentration, and conductance, the reciprocal of resistance, increases. To allow rapid comparison of results obtained with different instruments in different places, they are invariably expressed in terms of conductivity, which may be defined as the conductance of a solution between two parallel aligned electrodes 100 mm^2 in area positioned exactly 10 mm apart. Conductivity is generally expressed in mS (10 mm)$^{-1}$, a rather unorthodox unit in SI terms. This is exactly analogous to the more traditional units, mmho cm^{-1}, which are widely found in the earlier literature (Bower & Wilcox, 1965). Methods for the preparation of a saturation extract are well defined, and may be found in a comprehensive text on soil analysis (Rhoades, 1982b).

The susceptibility of plants to salinity damage depends upon the plant species and cultivar, and plant age, younger plants generally being particularly easily damaged. Production of high-yielding, salt-tolerant strains of cereals has been a major challenge to plant breeders over recent decades. General observations on plant response in soils with specified saturation extract conductivities are made in Table 4.2. They should only be regarded as a general guide.

Table 4.2. *Effects of salinity on the growth of plants*.

Description	Conductivity range mS (10 mm)$^{-1}$	Approximate % salt content	Effects[a]
Salt-free	0–2	<0.15	Generally negligible
Slightly saline	2–8	0.15–0.35	Restricted yields for many crops
Moderately saline	8–15	0.35–0.65	Only tolerant crops give useful yields
Strongly saline	>15	>0.65	Only a few very tolerant crops give useful yields

[a] The observed effect depends upon the species and the cultivar.

Under unfavourable climatic conditions, excessive salt concentrations may also occur as a consequence of fertiliser use. A quantitative comparison of the relative solubilities of fertilisers may be found in the salt index (see, for example, Jones (1982)). This assigns each material a numerical value relative to sodium nitrate, which is arbitrarily assigned a value of 100.

Sodic soils

The base cations adsorbed on the cation-exchange sites in a soil are displaced by hydrolysis. We should therefore regard soils as undergoing continuous exchange with the solution surrounding the soil colloids. Thus if a small amount of ^{45}Ca is added to a soil–water mix, the ratio of labelled to non-labelled calcium will eventually be similar for the exchange sites and the solution. Indeed, this approach is sometimes used experimentally to measure the amount of a particular cation which is exchangeable. Figure 4.10 represents what occurs when sodium or calcium is displaced. We have seen that, if sufficient calcium comes into solution at high pH, calcium carbonate precipitates. The same is true for magnesium, since dolomite ($CaMg(CO_3)_2$) is even less soluble than calcite. Reaction (4.15) liberates H^+ in passing from right to left, which would neutralise the hydroxide ion produced by the hydrolysis reaction (Fig. 4.10). Displacement of Na^+ by hydrolysis liberates hydroxide which remains in solution, thus raising the solution pH (remember that $K_w = \{H^+\}\{OH^-\} = 10^{-14}$).

Increases in soil pH above 8.5 are associated with Na^+ constituting

Fig. 4.10 Equations representing the hydrolytic removal of sodium or calcium from soil cation-exchange sites.

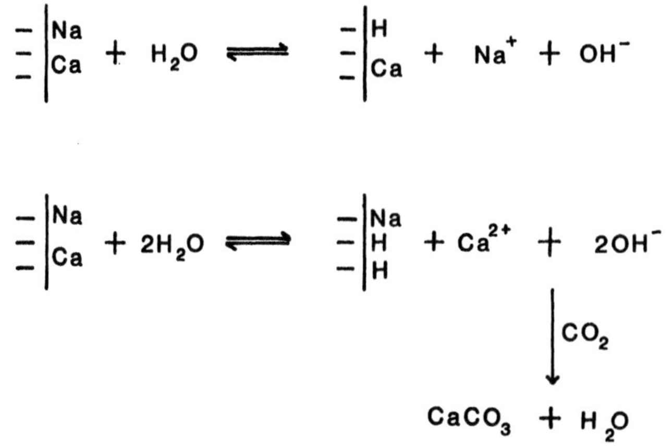

$$\begin{vmatrix} Na \\ Ca \end{vmatrix} + H_2O \rightleftharpoons \begin{vmatrix} H \\ Ca \end{vmatrix} + Na^+ + OH^-$$

$$\begin{vmatrix} Na \\ Ca \end{vmatrix} + 2H_2O \rightleftharpoons \begin{vmatrix} Na \\ H \\ H \end{vmatrix} + Ca^{2+} + 2OH^-$$

$$\downarrow CO_2$$

$$CaCO_3 + H_2O$$

>15% of the exchangeable cations in soil, but also with a low soluble neutral salt content (conductivity <4 mS $(10\text{ mm})^{-1}$). A high concentration of, say, sodium chloride in the soil solution suppresses the displacement of exchangeable Na^+ by H^+ (Fig. 4.10), as a consequence of the ratio law. A soil with a pH > 8.5, a saturation extract conductivity <4 mS $(10\text{ mm})^{-1}$ and Na^+ satisfying more that 15% of the CEC is known as a sodic soil. In strongly alkaline soils, the solubility of the soil organic matter increases (see Chapter 3). In arid climates this may lead to very dark deposits at the surface of the soil.

Saline-sodic soils

Between saline soils (saturation extract conductivity >4 mS $(10\text{ mm})^{-1}$; Na^+ <15% of CEC; pH < 8.5) and sodic soils come saline-sodic soils. These differ from saline soils in that >15% of the CEC is satisfied by Na^+. This additional category is a useful one because if such soils are irrigated and the neutral soluble salts are flushed out, they degenerate rapidly to sodic soils. Flushing neutral salts from saline soils by irrigation and drainage, on the other hand, improves the soil for plant growth because the pH will not be significantly increased.

Gypsum in soil

We should not leave the topic of soils in arid climates before discussing the role played by gypsum ($CaSO_4 \cdot 2H_2O$). The solubility of gypsum is defined by the equation (Lindsay, 1979):

$$CaSO_4 \cdot 2H_2O \rightleftharpoons Ca^{2+} + SO_4^{2-} + 2H_2O \qquad \log K = -4.41 \tag{4.19}$$

The solubility is not pH dependent (Fig. 4.9). If gypsum is added to a strongly alkaline, sodic soil, it will dissolve until calcite starts to precipitate, with the production of H^+ (equation (4.15)). The gypsum continues to dissolve, and the soil pH falls until (if enough gypsum is added), gypsum, and not calcite, limits the calcium activity in soil solution. Thus gypsum is a valuable product for the improvement of sodic soils.

Oxidation–reduction reactions in soils

So far in this chapter, we have not considered the changes in oxidation states of elements that may occur in soil systems. Those readers with a knowledge of chemical thermodynamics will be aware that the driving force in any chemical reaction is the minimisation of the free energy of the system. Equilibrium corresponds to the point at which the sum of the free energies of the products balances that of the reactants. In the case of mineral weathering, for example, the decrease in free energy associated with the formation of macromolecular crystalline lattices is such that further breakdown becomes energetically unfavourable. Indeed, it is sometimes possible to calculate equilibrium constants for reactions which are difficult to study directly from the additive properties of the free energies of formation of reactants and reaction products.

Consider a reversible reaction involving a reduction from an oxidised species (Ox) to a reduced species (Red):

$$Ox + ne^- + mH^+ \rightleftharpoons Red \qquad (4.20)$$

It may be shown (Glinski & Stepniewski, 1985) that the electrochemical potential for such a system, measured with a pure platinum electrode relative to a standard hydrogen electrode (or converted to that scale if an alternative reference electrode is used), Eh, is given by:

$$Eh = E_0 + 2.303RT[\log(\{Ox\}/\{Red\}) - mpH]/nF \qquad (4.21)$$

where E_0 is the reaction voltage when $\{Ox\} = \{Red\}$ and $\{H^+\} = 1$ mol l^{-1}, R is the universal gas constant, T the absolute temperature and F is the Faraday. The resulting equation is known as the Nernst equation. The Eh values for the major inorganic redox systems relevant to soils are listed for two system pH values in Table 4.3.

Table 4.3. *Eh values (mV) for important soil oxidation–reduction reactions at pH 5 and pH 7, after Russell (1987)*

	Eh (mV) at 25 °C	
Reaction[a]	pH 5	pH 7
Reduction of O_2	930	820
Reduction of NO_3^- to NO_2^-	530	420
Production of Mn^{2+} from MnO_2	640	410
Production of Fe^{2+} from $Fe(OH)_3$	170	−180
Reduction of SO_4^{2-} to S^{2-}	−70	−220
Reduction of CO_2 to CH_4	−120	−240
Production of H_2	−295	−413

[a] Equations for most of these reactions are given later ((4.22)–(4.27)).

It is appropriate, at this point, to consider exactly what the Eh values listed in Table 4.3 mean. Consider a soil which becomes waterlogged with static (non-flowing) water. The microorganisms present in the soil will start to use up the oxygen dissolved in the soil solution (see also Chapter 3). Some oxygen will diffuse down from the surface, dissolve at the water table, and then diffuse down through the solution. However, the rate of oxygen replenishment by this process is invariably much lower than the rate at which it is used up by the microflora, so the dissolved oxygen concentration declines. At this early stage, the Eh value is regulated by the concentration of the most easily reduced species, which is the residual oxygen. The appropriate version of equation (4.20) at this point then is:

$$O_2 + 4H^+ + 4e^- \rightleftharpoons 2H_2O \tag{4.22}$$

Equation (4.21) shows that *Eh* will fall logarithmically with the oxygen concentration. Eventually, the point will be reached where the extent of reduction of the next most easily reduced species becomes significant, in this instance, nitrate being reduced to nitrite. Many bacteria are capable of reducing nitrate to nitrite (Russell, 1987):

$$NO_3^- + 2H^+ + 2e^- \rightleftharpoons NO_2^- + H_2O \tag{4.23}$$

Note that, at this stage, the concentration of dissolved oxygen is still declining further, and reduction of the next most easily reduced species, manganese(IV) dioxide, has already started, but only at a very low level. The progression thus continues, with manganese(IV) dioxide, iron(III) hydroxide, sulphate and, finally, the hydrogen ion itself being reduced:

$$MnO_2 + 4H^+ + 2e^- \rightleftharpoons Mn^{2+} + 2H_2O \tag{4.24}$$

$$Fe(OH)_3 + 3H^+ + e^- \rightleftharpoons Fe^{2+} + 3H_2O \tag{4.25}$$

$$SO_4^{2-} + 10H^+ + 8e^- \rightleftharpoons H_2S + 4H_2O \tag{4.26}$$

$$2H^+ + 2e^- \rightleftharpoons H_2 \tag{4.27}$$

One disadvantage of *Eh* as an experimental parameter is that, for any redox pair, the value of *Eh* depends upon the pH, because H^+ ions are involved in the reduction. Moreover, the effect of pH differs for each reaction. Lindsay (1979) advocates the term (*pe* + pH) as a convenient single-term expression for defining the redox status of aqueous systems, where *pe* is defined as $-\log\{e^-\}$.

86 Soil chemical reactions

It is useful at this point to look at the merits of using $(pe + \text{pH})$ as a parameter to define the redox conditions prevailing in a soil. If both sides of equation (4.27) are multiplied by 0.5, the expression for the appropriate equilibrium constant becomes:

$$K = \{H_2\}^{0.5}/\{H^+\}\{e^-\} \qquad (4.28)$$

or:

$$\log K^0 = 0.5 \log \{H_2\} - \log \{H^+\} - \log \{e^-\} \qquad (4.29)$$

As mentioned earlier, Eh values are all expressed relative to the standard hydrogen electrode potential, which is defined as the potential generated when hydrogen gas at a pressure of 1 atm (101.3 kPa) equilibrates at a platinum surface with a solution with $\{H^+\} = 1 \text{ mol l}^{-1}$. To give a simple and consistent relationship between pe and Eh under these conditions, K^0 for equation (4.28) is assigned a value of unity. From equation (4.28), it follows that the electron activity, $\{e^-\}$, must also be arbitrarily unity for the standard hydrogen electrode.

When K^0 in equation (4.29) is assigned a value of unity, $\log K^0$ becomes zero, and equation (4.29) becomes, after rearrangement and using the p notation:

$$pe + \text{pH} = -0.5 \log \{H_2\} \qquad (4.30)$$

It follows from equation (4.30) that a $(pe + \text{pH})$ value of zero corresponds to equilibration with pure hydrogen gas at a pressure of 1 atm. For convenience of calculation, the older unit of pressure, the atmosphere [(atm); 1 atm = 101.3 kPa] has been retained here.

The equilibrium constant for equation (4.22) may be shown to be $10^{83.12}$ (Lindsay, 1979). Thus:

$$10^{83.12} = \{H_2O\}^2/\{O_2\}\{H^+\}^4\{e^-\}^4 \qquad (4.31)$$

from which, since $\{H_2O\}$ is close to unity:

$$4\text{pH} + 4pe - \log \{O_2\} = 83.12 \qquad (4.32)$$

or

$$pe + \text{pH} - 0.25\log \{O_2\} = 20.78 \qquad (4.33)$$

For pure oxygen at a pressure of 1 atm, $(pe + \text{pH}) = 20.78$. In soil, $\{O_2\}$ is unlikely to exceed ca 0.2 atm, because of the high nitrogen content of the atmosphere. Thus $(pe + \text{pH})$ is generally below 20.60, i.e. 20.78 + 0.25log 0.2. In practice, the oxygen content of the soil atmosphere is

Oxidation–reduction reactions in soils

likely to be less than 0.2 atm, because of microbial and root respiration. At any specified value of (pe + pH), the partial pressures of hydrogen and oxygen are fixed, and may be calculated from equations (4.30) and (4.33) respectively. Low soil oxygen contents are generally associated with increasing water content in the soil pores, because oxygen diffuses through water much more slowly than it diffuses through air (Russell, 1987; see also Chapter 3). Reducing conditions are most probable therefore in completely or partially waterlogged soils.

Figure 4.11 shows the link between pe and pH at various values of (pe + pH), and the corresponding values of oxygen and hydrogen partial pressures. The shaded zone of the figure represents the conditions likely to be encountered in soils (Lindsay, 1979). It should be stressed that the treatment of the (pe + pH) concept included here is a purely chemical one, based upon the approach described by Lindsay (1979). It takes no account of microbial reductions in soils which, in effect, generate higher than expected concentrations of reduced species in soil solution and create (pe + pH) and Eh values which are lower than predicted.

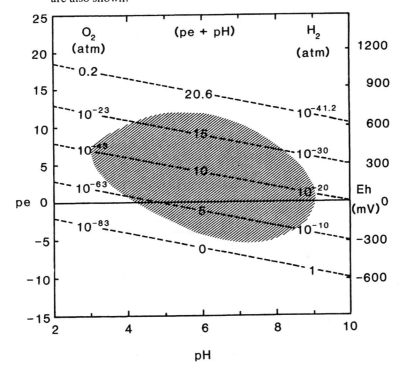

Fig. 4.11 The relationship between pe or Eh and pH at five different values of (pe + pH). The corresponding partial pressures of oxygen and hydrogen are also shown.

88 Soil chemical reactions

The relationship between Eh *and* (pe + *pH*)

Equation (4.21) may be rearranged to give:

$$Eh = E_0 + 2.303RT(\log [\{Ox\}\{H^+\}^m/\{Red\}])/nF \qquad (4.34)$$

The equilibrium equation for the general redox reaction, (4.20), may be written in the form:

$$\log K^0 = -\log [\{Ox\}\{H^+\}^m\{e^-\}^n/\{Red\}] \qquad (4.35)$$

or:

$$\log K^0 = -\log [\{Ox\}\{H^+\}^m/\{Red\}] - \log \{e^-\}^n \qquad (4.36)$$

Hence:

$$\log [\{Ox\}\{H^+\}^m/\{Red\}] = -\log \{e^-\}^n - \log K^0 \qquad (4.37)$$

Substituting the left-hand side of this equation into equation (4.34) gives:

$$Eh = E_0 + (2.303RT/nF)(-\log \{e^-\}^n - \log K^0) \qquad (4.38)$$

or:

$$Eh = E_0 - (2.303RT/nF)(\log K^0 - npe) \qquad (4.39)$$

By definition, *Eh* is zero for conditions corresponding to the standard hydrogen electrode, and $\{e^-\}$ is unity (i.e., $pe = 0$) for these conditions. It follows from equation (4.39) that $E - (2.303RT/nF)\log K^0$ must be equal to zero, and that:

$$Eh = (2.303RT/nF)npe = (2.303RT/F)pe = 59.2\, pe \qquad (4.40)$$

Values of *Eh* are also plotted in Fig. 4.11 for comparative purposes.

The solubility of iron and manganese under reducing conditions

The reduction of iron(III) hydrated oxide in soil is described by the equation (Lindsay, 1979):

$$Fe(OH)_3(soil) + 3H^+ + e^- \rightleftharpoons Fe^{2+} + 3H_2O \qquad (4.41)$$

for which $\log K^0 = 15.74$. Thus:

$$\{Fe^{2+}\}/(\{H^+\}^3\{e^-\}) = 10^{15.74} \qquad (4.42)$$

and:

$$\log \{Fe^{2+}\} = 15.74 - (pe + pH) - 2pH \tag{4.43}$$

Figure 4.12 shows how the activity of iron(II), $\{Fe^{2+}\}$, varies with pH for a range of values of $(pe + pH)$. It also shows how the activity of iron(III), $\{Fe^{3+}\}$, decreases with solution pH, based upon the equation:

$$Fe(OH)_3(soil) + 3H^+ \rightleftharpoons Fe^{3+} + 3H_2O \tag{4.44}$$

for which $\log K^0 = 2.70$. As $(pe + pH)$ decreases with increasingly reducing conditions, Fig. 4.12 shows that the solubility of iron increases.

The solubility of manganese also depends critically upon the redox conditions prevailing in the soil. Indeed, light rolling of soil prior to sowing seed is sometimes used as a method for prevention of manganese deficiency in cereals. The rolling must be sufficient to generate partial

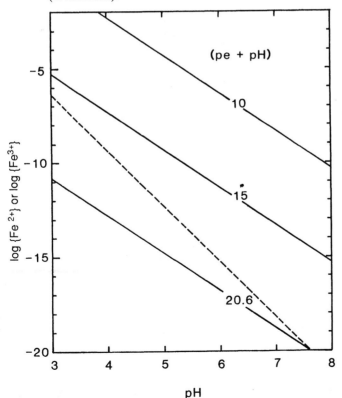

Fig. 4.12 The relationships between the activity of Fe^{2+} in solution and pH (solid lines) for three values of $(pe + pH)$, and between Fe^{3+} and pH (broken line).

anaerobiosis and reducing zones in the soil which increase manganese availability by reduction of MnO_2 to Mn^{2+}.

Reducing conditions and soil pH

The observant reader will have noticed that the reduction reactions occurring in soils and listed in equations (4.22)–(4.27) are all proton-consuming reactions. It might be expected, therefore, that creation of anaerobic conditions in soil by compaction and/or waterlogging should lead to an increase in the soil pH value, and this is what happens in practice when acid soils are flooded (Ponnamperuma, 1985). Flooding of sodic soils, and of soils with substantial amounts of free calcium carbonate, on the other hand, tends to cause a reduction in pH, due to the accumulation of carbon dioxide under the waterlogged conditions.

Chemical problems associated with anaerobic conditions

Prolonged exposure to waterlogged conditions may lead to a number of toxicity and other problems apart from the shortage of oxygen for root and microbial respiration. For example, iron and manganese may be solubilised in excessive quantities, which may prove toxic to plants. Toxic amounts of the sulphide may sometimes accumulate, although this problem tends to be offset somewhat by the precipitation of iron, manganese and zinc sulphides (Yu Tian-Ren, 1985). Zinc deficiency is a common problem in paddy soils, where strong adsorption as a consequence of the increased pH may contribute to the problem. Problems may also be encountered from organic toxins present as a consequence of the anaerobic fermentation of organic matter. The latter aspect was discussed in the previous chapter.

Sometimes problems may be secondary in nature. Rapid denitrification of nitrite produced even under mildly reducing conditions may lead to nitrogen deficiency occurring in the long term. The nitrogen is lost to the atmosphere as gaseous nitrogen or nitrous oxide (N_2O). This problem may be partially offset by fixation of gaseous nitrogen from the atmosphere by algae in the less reducing, surface zone of the flooded soil.

Biogeochemical cycling of nutrient elements

Cycles with no gaseous components

It should be clear by now that soil cannot be considered as a chemical system in isolation. It must always be thought of as a system which interacts continuously both with the atmosphere and with the biological populations that the soil supports. Superimposed upon this natural cycling is any effect which human activities such as agriculture, forestry, mining, waste disposal, water abstraction, atmospheric pollution and land reclamation, etc., might have. So far the examples of biogeochemical cycling that we have considered most fully are those of the base cations (see Fig. 4.4). In many respects, these are the simplest cycles, in that they do not have gaseous components as such. The only atmospheric inputs are dust, ash and other atmospheric particulates such as aerosols of maritime origin. Dust, erosion, fires, drainage water and plant and animal removal are the principal outputs. Other cycles which do not include gaseous components are those of iron, manganese, copper, zinc, cobalt, boron, molybdenum and phosphorus. Sketches representing the biogeochemical cycling of these elements would therefore all look very similar to Fig. 4.4.

Cycles with gaseous components

Cycles which involve gaseous components include those of hydrogen and oxygen (which are inextricably linked to the cycling of water in the gaseous and the vapour phases), carbon, nitrogen, chlorine, sulphur and selenium. It is worthwhile to consider briefly each of these in turn, to see what additional features are involved over and above those shown in Fig. 4.4.

Hydrogen

The cycling of hydrogen attracts little consideration from soil chemists, generally being considered only through its importance as a component of water, which is cycled through soil, often *via* rivers, to lakes or oceans, and hence back to the atmosphere. It is a major component element of biota and soil organic matter. Soil water contributes, after uptake by plants, to sugar formation *via* photosynthesis. Organic litter and plant leachates returned to the soil contain organic hydrogen which is recycled by the soil microbial population. Soil organic matter was considered in the previous chapter. Perturbations to the hydrogen cycle are not generally discussed separately from those of the hydrological (water) cycle, and are generally linked to climatic change of natural or pollution-driven origins.

The hydrogen ion is, of course, an important component of the hydrogen cycle, for reasons which have been discussed at length already. It need not be considered further here. The role of the hydrogen ion in water acidification will be examined in Chapter 6. Other minor gaseous components of the cycle include ammonia, ethane, ethylene, hydrogen sulphide and hydrogen gas itself. These species are of interest in their own right, but not as transport mechanisms for hydrogen on either a local or a global scale.

Oxygen

In common with hydrogen, the oxygen cycle is rarely considered on a global scale by soil chemists, although because of its importance to the oxidation–reduction reactions in soil, it is extensively investigated at a much more local level. Interest invariably centres around gaseous oxygen in the soil atmosphere, or dissolved oxygen in soil solution, the element's importance as a constituent of water generally being disregarded in the present context. It is an important component of soil and plant organic matter, and also, of course, of carbon dioxide, aspects which also are virtually ignored.

Ozone, O_3, attracts considerable attention in its own right as a potentially damaging pollutant and as an essential component of the upper atmosphere where it prevents transmission of excessive amounts of ultraviolet light. In both respects, it may ultimately interfere with the biogeochemical cycling of other elements. The impact of pollutants upon element cycling is a research area which is attracting considerable interest at the present time. Oxygen is also a component of sulphur dioxide, oxides of nitrogen and nitric acid vapour, each of which is significant as a pollutant rather than as a mode of transfer of oxygen. Oxygen in the oxyanions carbonate, bicarbonate, sulphate, nitrate, nitrite, phosphate, borate and molybdate is of little consequence in its own right.

Carbon

The carbon cycle is generally regarded as the one which is central to continued production and breakdown of all natural organic matter in any ecosystem, because of the function of carbon dioxide in photosynthesis and biological respiration. It was considered fully in Chapter 3. Carbon dioxide is vital to the buffering of the pH of many freshwaters (see Chapter 6) and to the regulation of climate. The concentration of carbon dioxide in the atmosphere passes through an annual cycle, but superimposed upon this at the present time in an almost exponential increase with time (see, for example, Skiba & Cresser (1988)). The increase is in part due to the greater use of fossil fuels, in part to the decline of the world's great forests (Bowman, 1990).

Generally environmental scientists do not regard the increase in atmospheric carbon dioxide as particularly disturbing, apart from the possible climatic change it might cause. It is correctly argued that the gas is often enriched deliberately in horticultural glasshouses to speed up growth or to regulate physiological growth stage. It should be remembered, however, that increased growth makes greater nutrient demands upon soils. These are unlikely to be a problem in cultivated, agricultural soils, but might be problematic in natural ecosystems. Rapidly induced changes in physiological growth stage could also lead to problems, for example, through changed frost hardiness or increased susceptibility to pest attack. These aspects require careful consideration.

Nitrogen

The nitrogen cycle has already been considered at various points in Chapters 1, 3 and 4. The element does not occur to any useful extent in primary minerals, and therefore inputs of nitrate or ammonium ions in precipitation or as dry deposition, or fixation of nitrogen gas, are an essential prerequisite to soil formation. We have also seen that nitrous oxide, N_2O, and nitrogen gas may be lost from anaerobic soils as reducing conditions develop. From alkaline soils, whether natural or a consequence of bad management, volatilisation of ammonia may occur (see also Chapter 6).

Ammonia and oxides of nitrogen from natural or pollution origins such as vehicle exhausts or industry may participate in a number of reactions in the atmosphere to yield nitric acid, an important component of acid precipitation. These are discussed in the useful text on atmospheric chemistry by Brimblecombe (1986). The significance of the end products and their interaction with the soil-plant system is discussed in Chapters 6 and 7. Some foliar surfaces or their associated microflora may take up nitrate and ammonium directly from intercepted precipitation. Figure 4.13 illustrates the essential components of the nitrogen cycle. A fuller description of the cycling of the element has been published by Freney and Galbally (1982).

Sulphur

The main gaseous component of the sulphur cycle, sulphur dioxide, has become a very emotive topic over recent years because it is so

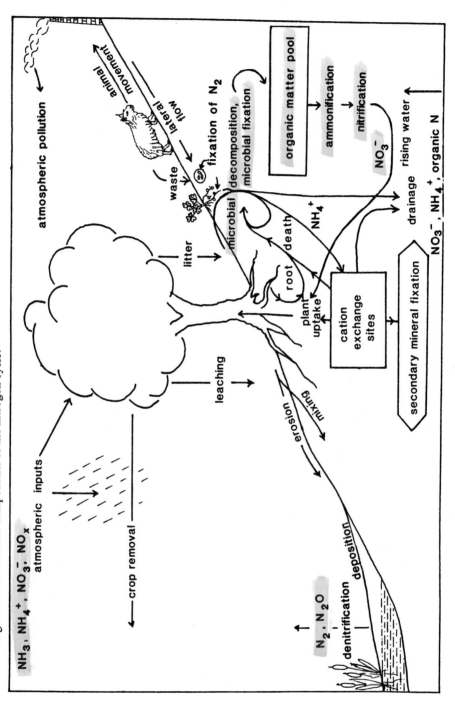

Fig. 4.13 The essential components of the nitrogen cycle.

inextricably linked to the 'acid-rain' debate (Cresser and Edwards, 1987). The major components of the sulphur cycle are illustrated in Fig. 4.14. The input of sulphur to the atmosphere as sulphate in oceanic spray sometimes is comparable (in regions with maritime climates) to that from pollution, and sulphate from both sources must be considered when investigating the fate of pollutant sulphur in soils (see Chapter 7). In this respect the sulphur cycle differs markedly from that of nitrogen. Sulphur is also transferred to the atmosphere from marshes as sulphur dioxide, especially in estuaries, and hydrogen sulphide, although the transfer is difficult to quantify. As mentioned earlier, sulphate may be strongly bound to hydrated oxides of iron and aluminium in acid soils. This is one reason why it is so difficult to build up a balanced sulphur cycle on a global scale (Marr & Cresser, 1983).

Chlorine

Biogeochemical cycling of chlorine is of little importance in the present context, at least as far as possible deficiencies are concerned. Plant requirements are generally small, and inputs from the oceans to the atmosphere and from the atmosphere to the soil are generally sufficient, and may be substantial. Hydrogen chloride is a product of industrial emissions, but has attracted much less attention than sulphur dioxide or oxides of nitrogen in discussions about acid precipitation. The most significant aspect of the chlorine cycle (and of the sodium cycle with which it is closely linked) is probably its contribution to the evolution of saline soils, which were discussed earlier in this chapter.

Selenium

The selenium cycle has not been studied to any great extent except on relatively local scales (Jacobs, 1989). This is surprising in view of its known importance in animal diet. Low molecular mass selenium compounds are volatilized from some plants.

Methods for studying element cycles

The discussion of element cycling so far has centred around processes rather than the methods by which individual components of cycles may be quantified. It is appropriate here to look briefly at what is involved, partly because it is an important branch of soil chemistry, and partly because it should help to reinforce the reader's general understanding of the cycling concept.

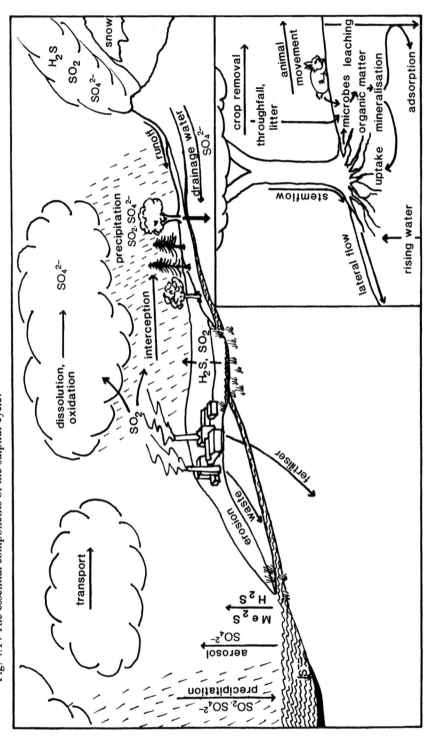

Fig. 4.14 The essential components of the sulphur cycle.

Precipitation inputs

Conceptually, at least, inputs in rain or snow are easiest to measure. A rain gauge must be designed which collects samples representative of a known area, without any possibility of contamination or adsorption losses of determinants. In practice, it is not easy to ensure that the presence of the gauge does not modify air currents to any significant extent, thus leading to collection of a non-representative volume of water. A high level of replication may be required, especially under a vegetation canopy, because of natural variability of both plants and soils. Collection of snow is particularly difficult, because of the possibility of drifting and resuspension. Without adequate forethought, site access and frozen water samples may also be a problem. Finally, open gauges trap a fraction of dry deposited material, as well as the wet deposition. Wet-only gauges are available, but they sometimes give poor reproducibility, and may require a power supply.

Dry deposition inputs

As far as the authors are aware, no method is yet available which allows satisfactory estimation of the dry deposition inputs of all species. For some ions, for example sodium and magnesium, in areas with substantial maritime-derived inputs, it may be possible to estimate dry deposition inputs from the excess chloride output from a catchment. This presupposes that no chloride is retained in the catchment on a year-to-year basis, that no chloride is derived from weathering, and that the ratio of sodium to magnesium to chlorine is similar for both wet and dry depositions. For gaseous inputs it is possible to calculate deposition rates by measuring concentration gradients from the soil or vegetation surface. An alternative approach is to enclose a branch in a suitable transparent container through which an artificial atmosphere is passed, and to measure the change in the composition of the atmosphere. This may be used to measure rates of carbon dioxide assimilation, for example.

Litter

Plant litter may be collected in suitable open-mesh bags suspended under trees or other vegetation. The open mesh allows some leaching of determinants, but this is preferable to the generation of an anaerobic compost. The system may, depending on the site, need to be sufficiently robust to withstand heavy snowfalls. It may, as indeed may all such apparatus, suffer from unintentional vandalism by any of a variety of animals (curiosity is a strong driving force!) or from intentional vandalism by humans.

Litter decomposition

Litter decomposition rates are often measured by inserting relatively small nylon mesh bags containing known amounts of appropriate litter into litter horizons for six months or longer, and then collecting them in for analysis.

Outputs in drainage waters

Drainage water may be sucked out of soil at the depth of interest by use of tension lysimeters. These essentially consist of porous ceramic or teflon cups attached to an impervious plastic tube. The cup is inserted at the relevant depth, and the tube is evacuated to the desired suction (tension), and left to suck water from the adjacent soil. Problems may be encountered from contamination, from freezing, or from poor contact with the soil.

Gaseous fluxes

Gaseous fluxes may be examined by sampling known volumes of air at different heights, or by allowing diffusion to transport the gas towards a suitable absorbent for a specified period.

Animal wastes

Transfers in animal wastes may be assessed from animals contained in suitable housings and fed artificially. In the field they may be assessed by attaching suitable receptacles to the animal. The potential problems don't need to be spelled out here!

Microbial cycling

Microbial respiration may be measured directly by using carbon dioxide traps. Microbial nitrogen transformations are best studied by the use of ^{15}N, a stable isotope of nitrogen. For example, if a labelled ammonium salt is added to a soil, and the soil incubated, it is possible then to extract the nitrate produced with potassium chloride solution and measure its ^{15}N content. This indicates the rate of conversion of labile ammonium-nitrogen to nitrate-nitrogen. Labelled fertilisers may be used in this way to study the fate of fertiliser nitrogen – how much is leached and in what forms, how much is trapped in the soil, how much is lost by denitrification, and how much ends up in the plant shoots and roots.

Rain simulation techniques

When researchers are interested in investigating effects of pollutants on element cycles, subjecting microcosms (cores removed from ecosystems complete with vegetation and with minimal disturbance) to simulated precipitation may prove very helpful. Collection of several similar cores allows replicates to be subjected to different pollutant loadings, thus facilitating the provision of reliable controls and subsequent interpretation of the results. Further discussion of simulation techniques may be found in a review by Skiba, Edwards, Peirson-Smith & Cresser (1987).

General comments

It is not possible in a soil chemistry text of this length to discuss further the methodology of quantification of element cycling. The brief discussion above should at least give the reader some idea of what is involved, and should help to clarify some of the statements about processes made throughout this book. The interested reader is encouraged to consult a recent text on the subject (e.g. Harrison, Ineson & Heal, 1990).

5

Soil fertility

What is soil fertility?

The term 'soil fertility' is difficult to define precisely. Generally it describes an overall soil condition which is the combined product of various natural processes and management options. A 'fertile soil' is one which is capable of producing a desired crop with favourable yield and quality characteristics. Large areas of the world still suffer from severe restrictions to agricultural productivity as a result of soil chemical, physical and/or biological limitations. The example of excessive salt build up has been mentioned previously (Chapter 4). It has been suggested that approximately 35% of the world's irrigated land suffers from suboptimal productivity (Carter, 1982).

For any production system to be sustainable, nutrient inputs and outputs must be balanced over a reasonable timescale. The nutritional requirements of individual plant species vary considerably and therefore the fertility of a soil varies for individual plant species (Brown & Jones, 1977). Removal of nutrient elements with the harvested crop is a major potential cause of decline in soil fertility, and losses should therefore be replenished. Typical values for the amounts of major nutrients removed by a range of crops are shown in Table 5.1. While in the present context chemical aspects of fertility are of prime concern, it is difficult to separate these from the closely linked physical and biological processes.

Natural levels of fertility are seldom adequate to sustain a reasonable long-term degree of agricultural productivity. It is therefore essential to be aware of soil nutrient reserves, the likely crop requirement and ways of counteracting a shortfall between the two. The aim of this chapter is to give a general overview of key chemical aspects of soil fertility, but with frequent reference to recent papers and reviews which should be consulted for deeper insight.

What influences natural fertility?

Numerous factors directly influence a soil's potential fertility. Of particular importance is the mineralogy of parent material. Many of the fundamental chemical and texture-related properties of mineral soil discussed in earlier chapters (including cation exchange capacity, base saturation and pH) may be attributed directly to the parent material and its secondary weathering products. Table 5.2, for example, shows the effect of parent material upon analytical data for various freely draining brown forest soils (cambisols) in NE Scotland. The Scottish Soil Survey

Table 5.1. *Typical rates of element removal for a range of crops (kg ha^{-1})*

Crop type	Yield[a]	N	P	K	Mg	Ca
Spring Barley						
Grain	5.52	56	12	20	6.7	2.1
Straw	3.58	12	1.8	38	2.5	6.8
Winter Wheat						
Grain	8.13	130	25	44	9.4	2.9
Straw	5.01	33	5.2	59	3.4	6.6
Potatoes	9.47	138	14	182	8.3	8.3
Conifer trees		21	2	6	3	14

[a] Dry matter (tonnes ha^{-1}).

Table 5.2. *Influence that parent material has upon some key soil chemical characteristics*

Association	Exchangeable cations (me/100 g)				%BS[a]	pH
	Ca	Mg	K	H		
Fraserburgh (shell sand)	20.8	1.21	0.04	nil	100	9.05
Leslie (serpentine/gabbro)	3.02	27.5	0.16	3.56	90.0	6.52
Insch (gabbro)	2.90	0.13	0.05	10.0	20.7	5.55
Countesswells (granite)	0.16	0.18	0.10	8.63	5.9	4.82

[a] Base saturation (from Glentworth and Muir, 1963).

groups soils for mapping purposes into 'Associations' and 'Series'. The former are based upon common parent materials, the latter upon drainage status. Soil parent material may also greatly influence the distribution and availability of trace elements. A survey across Northern Ireland by Dickson & Stevens (1983) showed a strong relationship between extractable soil copper and the underlying geology.

Several variables other than parent material also have important local influences on soil fertility. Reith *et al.* (1984) demonstrated the importance of soil rooting volume for the supply of adequate nutrients for crop production. Topsoils from eight soil series were collected and substituted for the local topsoil while keeping the original subsoil. Two different depths of soil were deposited. Crop yield showed a highly significant response to the increased depth of topsoil (Table 5.3). This was seen as good justification for the inclusion of this parameter in guidelines used for assessing land capability for agriculture (Bibby, Douglas, Thomasson & Robertson, 1982). An interesting discussion of how land evaluation systems are developed and their regional application has recently been published (MacDonald and Brklacich, 1992). Subsoil may also make important contributions to plant nutrient and water uptake, especially in situations where nutrient supply in the topsoil is poor (Kuhlmann & Baumgartel, 1991). Both the total amount and concentration of available nutrients in a soil profile are important with regards a soil's fertility status.

The mobility and extractability of many trace elements are affected by gleying, which may occur when soils become waterlogged (Berrow & Mitchell, 1980). In particular, drainage conditions may have a large effect on the amounts of cobalt, copper and molybdenum that can be extracted from cultivated soils (Berrow, Burridge & Reith, 1983). From the information on effects of redox potential presented in Chapter

Table 5.3. *Influence of topsoil depth upon dry matter yield for various crop types* ($Mg\ ha^{-1}\ y^{-1}$)

	Grain	Swedes	Grass
Topsoil (23 cm)	3.99	8.05	9.52
Topsoil (46 cm)	4.37	8.93	10.14
Standard error	0.14	0.10	0.15

After Reith *et al.* (1984).

4, iron and manganese especially should be solubilised in poorly drained soils as a consequence of reduction. However, trace elements that are not themselves reduced may be brought into solution by reduction of iron(III) oxides to which they are otherwise bound.

That soil has a very heterogeneous nature is a recurring theme throughout this chapter. It is important to highlight this non-uniformity of soil and to think about the limitations that it imposes upon attempts to understand, and therefore predict, a soil's fertility. A growing plant root will experience a diverse range of soil conditions (e.g. temperature, pressure, moisture, pH and nutrient status) as it develops and exploits new soil volumes. A recent text (Porter & Lawlor, 1991) on this subject is well worth reading. How plant root systems are capable of rapid adaptation to exploit localised nutrient-enriched soil patches has been discussed by Jackson, Manwaring & Caldwell (1990). This concept has important implications for the theory and modelling of nutrient uptake by plants.

Management options and soil fertility

Greatly improved control of soil fertility has played an important role in maximising agricultural production over recent decades. It is therefore worthwhile to consider the possible approaches to improvement of fertility that are widely used.

Crop rotation

At one time it was common practice to change the crops grown in a soil on a regular basis, with five, six or seven year cycles, a practice known as crop rotation. This minimised carryover of pests from year to year, and fallow periods gave the soil time to 'recover' from excessive removal of nutrient elements by crops in preceding years.

With the substantially increased use of fertilisers and agrochemicals, a crop rotation system has become unnecessary in many countries. This development has resulted in the possibility of continuous arable cropping, possibly even with two harvests a year under sufficiently favourable climatic conditions, and has allowed a greater degree of specialisation. The mixed farming system, where animal waste returns and crop rotation played essential roles in maintaining adequate soil nutrient status and productivity, is no longer necessary from the purely chemical viewpoint.

In Europe, various sequences of rotation, usually with local crop preferences and variable overall lengths, have been employed (Table 5.4) with great success, showing sustainability over long periods. The use of legumes such as clover in a grass ley, or a period of fallow to allow some degree of nutrient recovery, played a vital part. Overproduction and recent environmental concerns have renewed interest in these so-called reduced-input sustainable systems (McRae, Hill, Mehys & Henning, 1990).

Agriculturists in the tropics have developed their own systems to counteract the problems associated with nutrient-poor, often acidic soils and to meet the ever growing demand for food. For many tropical soils the fertility balance is delicate. This is particularly evident for rainforests (Proctor, 1987). The traditional practice of shifting cultivation relies on an initial vegetation burn which releases nutrients and often raises the soil pH. The pH rise is a consequence both of destruction of acidic organic components in soil and the production of ash with liming properties (e.g. carbonates). Lal & Ghaman (1989) noted an increase in exchangeable base cations in the top 300 mm as a result of burning rainforest. The associated increase in pH was also substantial, from 5.7 to 7.0 in the 0–100 mm layer and was still evident in the 200–300 mm layer. They also showed the effect that burning in windrows can have upon variations in soil fertility in an otherwise uniform field.

After the cropping period, fertility is restored by reverting to a fallow phase, the length of which depends upon the duration of cultivation (Lal, 1986). A ratio of one year arable to five years (at least) of fallow is desirable. During the fallow period, nitrogen may be increased by

Table 5.4. *Typical crop rotations of the UK (there is considerable local variation)*

Norfolk	NE Scotland
Grass/Hay	Hay
Wheat	Grass
Potatoes	Grass
Barley (undersown)[a]	Oats/Barley/Wheat
	Oats/Barley
	Turnips/Potatoes[b]
	Barley (undersown)[c]

[a] Lime applied.
[b] Limed for turnips only.
[c] Limed after potatoes, to minimise infection with streptomyces scabies (potato scab), which is favoured by high pH.

fixation, and nutrient elements resulting from biogeochemical weathering slowly start to build up in the soil. Conversion of forest to pasture can result in smaller losses of carbon and nitrogen compared to the losses when arable crops are grown. Brown & Lugo (1990), in a study in Puerto Rico and the US Virgin Islands, showed that recovery of soil carbon concentration during succession took 2–3 times longer than that of soil nitrogen concentration.

Intercropping

Diverse mixtures of plant species have been chosen by farmers over the centuries to ensure best possible use of native soil fertility and rainfall. There are numerous approaches to these intercropping systems, which often involve a legume (perhaps a multipurpose woody species) together with a food or forage crop (Kang, Wilson & Spiken, 1981). The overall effect of using such a system tends to be a reduction in the length of fallow period required. The more efficient nutrient use in mixed vegetation systems is a consequence either of a larger soil mass being exploited or the same soil being exploited more completely. An advantage of deep rooting species is therefore related to their increased potential to cycle nutrients and water from the subsoil. There may also be a difference in the timing of peak demand for nutrients by each species.

While there are many advantages, some disadvantages also occur with mixed systems through the direct competition for any potentially limiting resources. This occurs, for example, in pastures where legumes are present. Legumes tend to be poor competitors with grass for phosphorus, potassium and sulphur. If a deficient pasture is fertilised with phosphorus, a change in species composition (greater proportion of legumes) may occur, as well as an increase in dry matter production.

Unavailable nutrient pools

Many soils, while containing substantial amounts of nutrients, suffer from the lack of their availability. The highly organic upland soils of NW Europe are a good example. Typical values for nitrogen, phosphorus and potassium in the 0–100 mm layer of a deep peat are 2238, 110 and 50 kg ha^{-1} (Williams, personal communication). However, these major plant nutrients are largely bound up in organic matter and microbial biomass. In these situations it is essential to improve the rate of turnover and cycling of nutrients if the soil is to be used for cropping.

106   Soil fertility

Plant breeding and selection

It is often difficult to separate the individual contributions made to increasing crop yields by plant breeding and selection from those achieved by improving fertiliser use efficiency and soil nutrient status. The importance of plant breeding and selection has, however, been demonstrated by Riggs *et al.* (1981) in their comparison of yields of 37 old and new barley varieties used between 1880 and 1980 in England and Wales. The average genetic gain in yield was 0.39% per year over the entire 100-year period, but 0.84% per year between 1953 and 1980. Similar information is also available for winter wheat (Austin *et al.*, 1978). However, exploitation of higher yielding crop varieties or genotypes always results in greater nutrient demands being put upon the soil. The success of recent plant breeding and selection trials for tolerance to acid soils (usually to high aluminium) has been particularly impressive (Taylor, 1988).

Use of fertilisers

The use of fertilisers has become an essential and routine part of many farming systems. Applications of nitrogen, phosphorus and potassium make up the majority of the market. The effects of various combinations of nitrogen, phosphorus, potassium fertiliser on herbage composition and yield for an established grass/clover sward are shown in Table 5.5. While the addition of a single nutrient (nitrogen, phosphorus or potassium) does increase yield, a combined application (nitrogen + phosphorus + potassium) causes more than just a cumulative increase. Herbage composition also closely reflect the fertiliser applied.

Table 5.5. *Influence of various fertiliser combinations on the composition (%) and dry matter yield (Mg ha^{-1}) for a grass clover sward*

Fertiliser[a]				Composition		
N	P	K	Yield	N	P	K
0	0	0	3.6	1.63	0.18	0.66
0	0	1	5.0	1.67	0.17	2.23
0	1	0	4.4	1.84	0.33	1.08
1	0	0	5.8	2.08	0.16	0.39
1	1	1	11.0	1.97	0.33	2.17

[a] 1 denotes fertiliser applied, 0 denotes no fertiliser.

It is possible to purchase a wide range of fertilisers containing a single nutrient element (e.g. ammonium nitrate) or compound (mixed) forms in which two, three or more nutrients are present. Application may involve either solid broadcasting, liquid injection or irrigation. Trace elements are applied in a variety of ways which include seed coatings, liquid and solid forms. For soils where an added trace nutrient may be rapidly fixed, it may be better to apply the nutrient *via* a foliar feed in dilute solution form. This is done, for example, to rectify manganese deficiency on soils with excessively high pH values.

Compound fertilisers are produced in two distinct ways, either as chemically distinct forms (i.e. true *chemical compounds*, such as potassium nitrate) or by mixing the individual chemicals together. Compound fertilisers containing various proportions of nitrogen, phosphorus and potassium are available to suit the majority of crop types and situations. The proportion of the total fertiliser applied as a compound fertiliser for various regions of the world is shown in Table 5.6. The advantages of compound fertilisers lie in the need for only a single application which consequently reduces labour costs and minimises soil structural deterioration through unecessary trafficking.

Modern industrially produced fertilisers tend to be relatively pure and largely water soluble. Nitrogen fertilisers are produced by various industrial processes through the chemical combination of atmospheric nitrogen with either oxygen (nitrate) or hydrogen (ammonia). These processes require considerable inputs of energy but the expense of production is justified by the resulting yield responses.

Large deposits of potassium salts are available world wide, the salt often having been formed as a result of evaporation from an isolated water body. Over 90% of the world's total ore production goes to agriculture, but some degree of purification is often required. The

Table 5.6. *Proportion (% by weight) of the total fertiliser applied as a compound (all data for 1987/8 period)*

Nutrient	World	'Developed' countries	'Developing' countries
N	18.2	27.4	17.2
P	62.6	81.0	59.1
K	36.4	57.3	33.0

Calculated from Fertiliser Product Consumption Forecasts (1989).

sources of native phosphorus deposits tend to vary considerably in chemical composition and phosphorus concentration. The mined material may undergo a wide variety of processing steps, either to alter chemical form to increase its solubility or to increase the relative phosphorus concentration. The low solubility of many rock phosphates means that their direct application to soil tends to be limited to specific situations, such as acidic soils where dissolution rates may be enhanced.

Nitrogen fertilisation

Nitrogen is the key nutrient responsible for much of the recent increase in agricultural production. Nitrogen cycling is mainly regulated by biotic processes, with microorganisms playing a central role in controlling the availability of soil nitrogen to plants and the fate of nitrogen fertiliser (see also Chapter 3). Both these factors are therefore highly dependent on climatic variables such as temperature and moisture content. Less than 3% of the total soil nitrogen is converted annually to soluble forms (see Chapter 3). Nitrogenous fertilisers are used in extremely large quantities by both developed and developing countries. The dominant form of nitrogen applied has changed over the last 20 years (see, for example, Table 5.7 for the UK). In 1980/1, Rosswall and Paustian (1984) estimated that the 60 Tg yr^{-1} of fertiliser nitrogen applied globally represented 40% of that fixed biologically in all terrestrial systems and was 36% more than was fixed in cropland. This figure must be viewed against the fate of applied and natural nitrogen, and Table 5.8 shows estimates of major global losses of nitrogen from agro-ecosystems. The need for the high fertiliser nitrogen inputs is then much more readily apparent.

The total amount of nitrogen in a profile often tends to decrease as a

Table 5.7. *Change in the types of straight nitrogen fertiliser applied in the UK over the last 16 years (values expressed as a percentage of the total nitrogen applied).*

Year	Urea	Ammonium nitrate	Calcium ammonium nitrate	N solution
1973/4	–	70	15.1	11.0
1987/8	12	72.2	6.7	8.0

Calculated from Fertiliser Product Consumption Forecasts (1989).

result of cultivation. Smith & Young (1975), for example, noted that for eight major US soils used agriculturally, the total nitrogen content of virgin surface horizons was *ca* a third more than that of their nitrogen-fertilised, cultivated counterparts. The environmental and health concerns over intensive agriculture tend to be not that there is too much nitrogen but that it is sometimes present in the wrong forms.

The economic returns from efficient nitrogen fertiliser use are very favourable, helping to explain not only why relatively large amounts of nitrogen are applied but also why possible controls on nitrogen use are difficult to implement. For the UK, average responses from many fields and years show the yield of wheat increasing by 24 kg of grain with every kg of nitrogen fertiliser applied until the stage where the nitrogen versus yield response curve begins to reach a plateau. The resulting increased grain yield is worth approximately eight times the cost of the applied nitrogen (ignoring labour costs).

One of the main problems relating to nitrogen fertiliser use is a lack, in many parts of the world, of suitable advisory soil test methods for evaluating nitrogen fertiliser requirements. This, in part, reflects the important role biologically controlled processes play in the mobilisation and immobilisation of the substantial reserves of soil nitrogen (Goss, Williams & Howse, 1991). Table 5.9 shows estimates (Royal Society, 1983) of the total nitrogen content (a large proportion of which will be in an organic form) of UK farm and forest soils. The specific contribution these massive reserves make towards supplementing any fertiliser additions will show annual variations reflecting regional and local climatic differences. In consequence considerable emphasis must be placed upon such factors as previous cropping history, yield and fertiliser use, all of which reduce the likely precision of predictions made of fertiliser requirement. It is therefore clear that an understand-

Table 5.8. *Estimated global losses (Tg) of nitrogen from agro-ecosystems (from Rosswall and Paustian (1984)).*

Harvested products	30
Leaching losses	2
Erosion	2–20
Denitrification	1–44
Ammonia volatilisation	13–23

ing of factors influencing the interaction between individual components of the nitrogen cycle is essential.

There is considerable concern over potential risks to public health caused by increasing nitrate concentrations in both ground (e.g. Madison & Brunett, 1984) and surface waters (ECETOC, 1988). More than 30 million people in America, for example, still have private water supplies, and an additional 74 million use public water supplies served by groundwater sources (Lee & Neilsen, 1987). The increase in amounts of leachable nitrate in soil is in part due to a mismatching of the soil's capacity to supply and the crop's potential to take up nitrogen.

It is easy to conclude that the current situation is a direct result of the tremendous post-1945 global increase in nitrogen fertiliser use which, for the USA, was 11 fold between 1950 and 1980 (Lee & Neilsen, 1987). Crop uptake efficiencies calculated using ^{15}N (the heavy nitrogen stable isotope) labelled fertiliser have demonstrated efficiency values of between 40% and 80%. The appropriate value within this range varies considerably with plant species, arable crops, for example, generally showing lower efficiencies than grass (Smith, Elmes, Howard & Franklin, 1984). The reactions and interactions involving fertiliser nitrogen in soil, however, are complicated. Except under specific conditions (e.g. heavy rainfall immediately after application), nitrogen fertiliser added early in the growing season has been shown to be present at relatively low extractable amounts in soil at the end of the season. This suggests that the current year's fertiliser makes a limited direct contribution to autumn leaching losses (Fig. 5.1). The nitrate in crops therefore primarily has been mineralised from soil organic nitrogen (Goulding *et al.*, 1989). It has, however, been clearly shown using data from Roth-

Table 5.9. *Estimates of the total nitrogen contents of UK farm and forest soils* (*based upon data from Royal Society Report* (1983))

	Area (Mha)	Nitrogen in soil (Mg ha^{-1}) to depth of 35 cm	Total N (Tg)
Arable grass <5 years	7.0	7.0	49
All grass >5 years	5.1	12.5	64
Rough grazing	6.2	20.5	127
Forestry	2.4	20.5	49

amsted's long-term Broadbalk Experiment that the plots receiving repeated nitrogen applications of 144 kg ha^{-1} annually since 1852 contain 20% more organic nitrogen than those on the unfertilised plot (AFRC Institute of Arable Crops Research, 1989).

Phosphorus fertilisation

Maintaining an adequate soil phosphorus status is complicated by the many mechanisms by which the element may be fixed, especially in acidic soils (Chapter 4). Benefits arise not only directly but also from improved crop establishment. Stimulated root growth can be observed in areas of soil fertilised with phosphorus compared to non-fertilised areas (Anghinoni & Barber, 1980). A massive amount of literature exists, covering all aspects of soil phosphate chemistry (e.g. Sanyal &

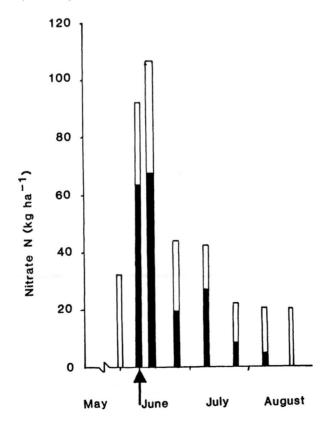

Fig. 5.1 Change in the amounts of soil (open) and fertiliser (shaded) derived nitrate nitrogen extractable with 2 M KCl for spring barley. Arrow shows when ^{15}N-labelled ammonium nitrate was applied at 125 kg nitrogen ha^{-1} (Macaulay Institute for Soil Research, 1986 Annual Report No. 56).

De Datta, 1991). However, many apparently conflicting reports can be found. This reflects the complex nature of phosphate reactions and interactions, and the great dependence upon local soil and plant factors.

Only a relatively small proportion of any phosphorus applied (<20%) is removed in the current year's crop, the remainder of the crop phosphorus being supplied from 'native' soil reserves. Predicting reliably the likely availability of these often substantial reserves is a difficult task. Benefits from large, one-off fertiliser applications may still be observed many years later. On the other hand, it is sometimes even possible to show early crop responses to fertiliser phosphorus on soils with a high phosphorus status.

Soil fertility classifications generally cover broad class ranges, often between 3–7 groupings for European countries (Hanotiaux & Vanoverstraeten, 1990). The central class is usually considered as being optimum. Once this moderate status is achieved, maintenance fertiliser applications are advised and calculated by estimating the amount removed by the crop and adding extra (e.g. using 1.2 times estimated crop offtake) to counteract any soil fixation that may have occurred. Recent trends in phosphorus status for various catagories of agricultural soil for England and Wales are available (Skinner, Church & Kershaw, 1992).

A wide range of extractants is currently used to quantify soil phosphorus status (Sanyal & De Datta, 1991). They vary greatly in their chemical 'aggressiveness' and range from water to acid or alkali. Isotopic exchange (i.e. using ^{32}P) has also been used extensively and can provide valuble information about relatively mobile, and hence plant available, phosphorus (Larsen, 1967). Ion exchange resins appear to offer a particular advantage for phosphorus studies by acting as a sink, which maintains a low phosphorus concentration in the extracting solution (Bache & Ireland, 1980). The wide range of procedures employed often makes comparison between regions and countries difficult. It is possible that a mixed anion/cation resin system may offer a universal and widely acceptable multielement method (Somasiri & Edwards, 1992).

Potassium fertilisation

Potassium is similar to phosphorus with regards to the large potential reserves present in many soils. The potassium content of mineral soils, for example, can vary between 0.04 and 3% (equivalent to 1000 to 75 000 kg ha^{-1}). Concentrations of soil solution potassium are controlled by the equilibria and kinetics of reactions between the various forms of soil potassium, the soil solution, and the divalent ion content in solution and on the exchange phase (Sparks & Huang, 1985). Exchangeable soil potassium generally represents a small proportion of the total potassium present (see Table 5.10).

A comparison of effects from a single large application of potassium with those of much lower annual applications or no potassium application at all has been made for an established grass sward. The single potassium addition (332 kg ha^{-1}) was five times the annual potassium dressing (66 kg ha^{-1}). Soil potassium concentration fell dramatically following the single application treatment (Fig. 5.2(a)), and after only two years it was similar to the concentrations for the plots receiving the much lower annual or zero treatments. Potassium removal in the harvested crop showed a similar trend to that of soil potassium (Fig. 5.2(b)). Large initial offtakes from the once-only application were followed by a sharp decline.

It is worth being aware of the possibility of element imbalances in soil or vegetation caused by either natural soil processes or fertiliser additions. The interaction between potassium and magnesium is especially well documented with regards to herbage quality and magnesium imbalances in the grazing animal (hypomagnesaemia). Competition effects on ion uptake can result from a combination of external or internal (to the plant) factors, i.e. absorptive or translocational factors (Barber, 1984).

Soil acidity

There remain large areas of potentially productive land which suffer limitations due to soil acidity related problems. Acid soil is

Table 5.10. *Exchangeable soil potassium as a percentage of the total potassium present in a soil from Delaware (based upon Sadusky, Sparks, Noll & Hendricks (1987)). The soil was rich in potassium feldspars*

Soil depth (mm)	Exchangeable	1 M HNO_3-extractable	Mineral[a]
0–230	0.7	1.2	98.1
850–1180	0.5	1.1	98.4

[a] Calculated by difference between total potassium and exchangeable plus HNO_3-extractable potassium.

114   Soil fertility

particularly widespread in tropical areas of Africa, America and Asia where acid soils cover 27, 57 and 38% of the total land area respectively.

Soil pH and its measurement

The determination of soil pH is probably the most common and apparently simple test performed on soil. Establishing the true pH of a soil sample is, however, beset by a whole range of analytical problems, as mentioned in Chapter 4, making a precise interpretation of agronomic significance extremely difficult. Soil acidification is a natural process

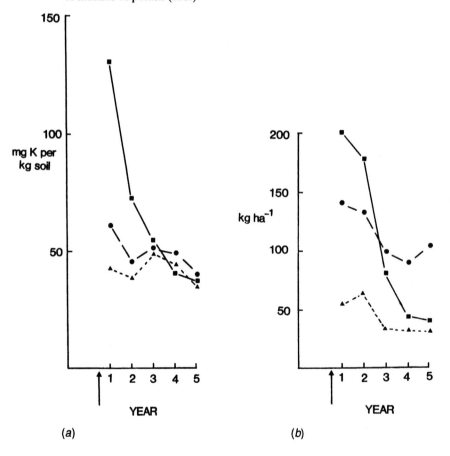

Fig. 5.2 Effects over a five-year period on acetic acid-extractable soil potassium (*a*) and plant potassium (*b*) of a single application of potassium at either 332 (squares), 66 (circles) or 0 (triangle) kg ha^{-1}. Potassium applied as muriate of potash (KCl).

(see Chapters 2 and 4) which results from processes such as plant uptake and leaching of base cations in drainage water. Modern agricultural practices can therefore cause an increase in acidification rates. An example of this has been reported by Haynes (1990), who compared the effect of nitrogen fertiliser type applied by trickle emitters, and pH in the wetted soil volume. Depth of soil acidification was directly dependent upon the relative mobility of the dominant nitrogen species and reached double the depth with urea fertiliser compared to ammonium sulphate. This deeper acidification was much more difficult to ameliorate.

As stated in Chapter 4, the pH value of a soil cannot be as precisely defined as that of a solution because it is the pH of a soil–water equilibrium system that is always measured. As a consequence the value obtained during measurement of soil pH is known to depend upon a number of experimental variables. These include the soil-to-solution ratio, the type and concentration of electrolyte used, electrode position and whether the suspension is stirred or not during the measurement (Bache, 1988).

It is generally accepted that more reproducible results can be obtained by using a weak (0.01 or 0.005 M) equilibrating soil solution, commonly either calcium or potassium chloride, rather than distilled water. While there is some debate as to the concentration that should be used, the aim is to mimic soil solution concentrations. This reduces the ion dilution effects which would otherwise result from adding distilled water only and reduces variability for a given soil. For acid soils the suspension pH will show a general decline with increasing salt content of the equilibrating solution. The difference for upland acidic soils of Northern Europe between the pH in 0.01 M $CaCl_2$ and that in water (both at a 1:2.5 soil:solution ratio) is often greater than half a pH unit.

Measurement of soil pH alone does not provide sufficient data for direct assessment of the lime requirement. This is due to the fact that pH measurement estimates only the activity of hydrogen ions in solution (Bache, 1988). No knowledge is forthcoming from a simple pH measurement of the relationship this intensity factor has with the total acidity of the system (i.e. the nature and amount of the proton donors in the solid phase). The latter, however, regulates a soil's response to lime applications (the buffer capacity, which is referred to again later in this chapter). Any assessment of lime requirement must take account of the soil's buffer capacity. In the normal pH range of 4.5–6.5 of agricultural soils (containing substantial amounts of organic matter), the pH-dependent charge associated with organic matter is generally the major buffering mechanism. Texture and organic matter content must therefore be taken into account when estimating the lime requirement from soil pH measurements.

Liming material

The effectiveness of the various types of liming materials depends upon their potential neutralising capacity and, because of their slow reaction rates, particle size. It should be remembered that the dissolution of calcite is slow, so that if sufficient ground limestone is added to a soil to raise pH to a desired value, it may be months before this pH is reached. Dolomite is less soluble than calcite for a given size distribution, and takes even longer to react. The solubility of carbonates is pH dependent, so that when pH rises sharply around a limestone particle, its dissolution rate is reduced. If the soil is well mixed by soil animals this is not a problem. In deep acid organic soils, however, where such mixing plays little part, sharp pH gradients may be observed over several years after limestone is applied to the surface only. A comparison of the effects on soil pH and levels of calcium and aluminium saturation through the surface application of gypsum or of deep lime incorporation to ameliorate subsoil acidity in a Typic Hapludult (Sumner & Carter, 1988) is shown in Figure 5.3. A marked reduction in soluble aluminium and an increase in soluble calcium occurred throughout the mixed profile. This was associated with deeper root penetration compared with that in the control.

Unlike the majority of fertilisers, lime is not applied annually, but perhaps every 4–5 years. This makes it essential that any liming operation be part of a well-balanced overall approach. If a rotational cropping system is used, lime applications should be timed to give maximum benefits to the most acid-sensitive crops.

Acidity and plant growth

Soil acidity tends to influence plant growth through indirect effects upon element availability and mobility rather than through direct effects of H^+ *per se*. The concentrations of many elements in soil solution are directly controlled by pH (see Chapter 4). The possible beneficial effects of reduced soil pH (increased trace element availability) are not generally sufficient to counteract the associated dramati-

Soil acidity

cally increased concentrations of toxic elements such as aluminium and manganese.

It is now readily apparent that plant species and genotypes within species differ widely in their tolerance to excess aluminium (Foy, 1988). The data in Table 5.11 illustrate the effect of soil acidity on the yield of various crop species growing on a mineral soil (iron humus podzol or typic fragiothod). The sensitivity of barley is especially apparent in this table; winter wheat showed no significant yield loss even at pH 5.2. A more detailed data set is shown in Fig. 5.4(*a*). The critical soil pH value (Bache and Ross, 1991) below which substantial yield reductions occur (in this case about 5.7) is dependent upon a number of variables which

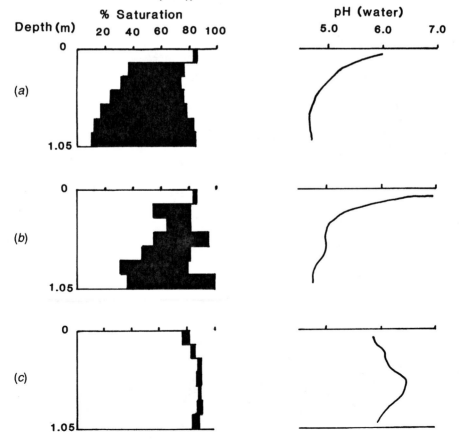

Fig. 5.3 Effect upon soil aluminium (shaded) and soil calcium (unshaded) % saturation, and upon soil pH, all versus depth in profile, (*a*) of no treatment (control plots); (*b*) of 10 tonnes gypsum ha^{-1}, mixed into topsoil, and (*c*) of 18 tonnes fine limestone ha^{-1}, mixed to 1 m depth (based upon data from Sumner and Carter (1988)).

118   Soil fertility

include plant species, nutrient status and type of soil. The once only application of large amounts of phosphate (2000 kg ha^{-1}) reduced the critical pH (Fig. 5.4(b)) from about 5.6 (low phosphorus) to 5.2 (high phosphorus).

The response of individual soil types to additions of lime (Haynes, 1982) varies considerably as a consequence of differing organic matter concentrations and clay mineral assemblages. In a comparison of two acid soils, Bache and Crooke (1981) found that the apparent critical pH

Table 5.11. *Influence of soil pH (water) upon grain yield and herbage production (expressed as a percentage of the maximum yield)*

Crop	Soil pH					Standard error
	4.8	5.2	5.6	6.0	6.4	
Spring barley	21.3	59.4	88.4	91.3	100	3.85
Winter wheat	74.2	98.2	100	100	95.5	3.69
Oilseed rape	72.0	84.8	91.5	93.8	100	2.14
Grass (3 cuts)	82.0	92.9	96.2	100	100	2.56

Fig. 5.4 Effects of soil pH on spring barley yield (*a*) with no added phosphorus and (*b*) with (filled circles) and without (open circles) phosphorus fertiliser (based upon data in Macaulay Institute for Soil Research, Annual Report No 54 (1984)).

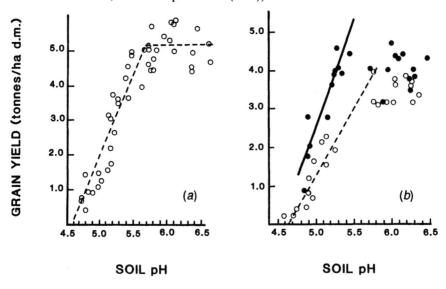

for the growth of barley in pots was 0.25 units lower in a soil showing lower aluminium solubility.

It is very difficult to apportion the effects of low soil pH on plant growth to either toxicity (aluminium or managanese) or to a specific nutrient deficiency. Aluminium toxicity, for example, is usually characterised by an inhibition of phosphorus uptake and translocation (Haynes, 1982), with the immobilisation of phosphorus on and in plant roots (Foy, Chaney & White, 1978). High concentrations of aluminium in soil solution cause severely restricted root extension and thus limit the degree of soil exploration and potential supply of nutrients, often resulting in the development of phosphorus deficiency symptoms. Maintaining an adequate soil phosphorus status often appears therefore to reduce the direct effects of soil acidity upon plant growth (Fig. 5.4(*b*)). The influence of soil acidity on the productivity of a wide range of crops has been discussed by Couto (1982).

The primary aim of adding lime is to alleviate the influence of toxic elements through raising a soil's pH. Additional and less obvious factors include the reduced effectiveness of some herbicides (Fox and Hoffman, 1981, see Chapter 7), which can increase the competition from weed species. The picture is therefore rather complicated, making it difficult to evaluate the individual effects when studying lime–pH–crop response patterns. The reported nutritional consequences resulting from liming are numerous, but are sometimes conflicting and usually also intimately associated with changes in plant growth patterns, such as changes in the distribution of roots.

Trends in soil pH

In Britain, general concern over levels of agricultural production during the 1930s prompted Government action to improve efficiency. This included a subsidy for lime. Over the 40 year life of this subsidy, average lime use increased by 10 fold to an average of 5 or more million tonnes per annum. This figure has now dropped somewhat since the subsidy was removed (Agricultural Lime Producers' Council, 1983).

Attempts have been made to estimate national figures for annual maintenance relative to the amount required to keep soil (<12% organic matter) at recommended pH values shown in Table 5.12. The figure for Scotland was calculated as being just below 0.5 million tonnes (calcium oxide equivalent) per annum (or 0.3 tonnes ha^{-1}), which matched the amounts applied during the sixties and seventies (Reith, 1980). Any decline in the amount of lime applied may not become apparent in soil pH trends of advisory samples for a number of years because of the slow reaction rate of lime. A decline would also not be expected to occur across the whole spectrum of agricultural land. Since 1969, the UK Agricultural Development and Advisory Service (ADAS) has carried out a soil sampling survey of arable and grassland fields in England and Wales, with the aim of obtaining unbiased estimates of soil nutrient status and establishing their trends with time (Church & Skinner, 1986). Results from this survey (Skinner et al., 1992) suggest that 30% of rotationally cropped land had pH values below the ADAS recommended value of 6.5 (water) and nearly half the continuous grassland was below pH 6.0 (water). It was noted in particular that 30% of barley was grown at a pH less than 6.5 and 10% below 6.0, where substantial yield reductions would be expected from the data in Table 5.11. The overall figure showed a strong regional distribution, with the greatest proportion of low pH soils associated with less agriculturally productive, predominantly grassland areas of Wales and Northern England.

There is considerable information available regarding the long-term loss of lime from agricultural soils. The main problem, however, which is not restricted only to liming experiments, relates to the extrapolation of these often very detailed local data to the wider context for soil types, climate, etc. Losses of calcium will occur primarily through leaching, with lesser amounts being removed in the harvested crops (see Table 5.1). Gasser (1973) and Bolton (1972) calculated annual losses of the order of 135–270 kg $CaCO_3$ ha^{-1} from a loamy sand of pH 5.8 (i.e., 54–108 kg Ca ha^{-1}). It also appears that the rate of calcium leaching increases as the target soil pH rises, doubling for each increase of one

Table 5.12. *The significance of soil pH (water) values for various crops*

pH range	Crop and soil comments
<5.0	Possible failure of all crops
5.0–5.4	Possible failure of barley, oilseed rape, peas and beans
5.5–5.9	More sensitive crops may suffer from acidity symptoms, which may be evident in low pH patches
6.0–6.5	Suitable for most arable crops. Trace element problems may arise at the higher values
>6.5	Trace element deficiences likely on many soils

Source: SAC/MISR (1985).

pH unit. The difference in annual calcium leaching between maintaining a soil pH of 6.0 or 6.5 was 235 compared to 336 kg $CaCO_3$ ha^{-1} (Gasser, 1973).

Additional factors which can cause a rapid increase in soil acidity include the application of certain nitrogen fertilisers (Fig. 5.5). It can be calculated that theoretically, with ammonia (either anhydrous ammonia, urea or ammonium nitrate), 3.57 kg $CaCO_3$ is required to counteract the acid generated by each 1 kg of nitrogen which is nitrified. Observed values tend to be much lower in practice, with difficulties introduced as a result of soil mixing caused by tillage operations (Jacobsen and Westerman, 1991).

Effects of liming on soil solution chemistry

Apart from the obvious effects of liming, which include increasing pH and exchangeable calcium concentrations, various other changes are likely. Curtin and Smillie (1986) noted, using laboratory

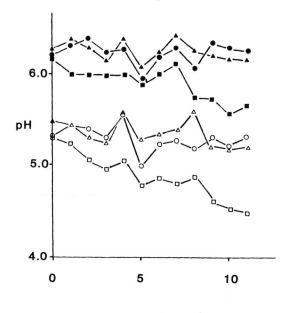

Fig. 5.5 Influence on soil pH under grass of applying fertiliser nitrogen at 90 kg N ha^{-1} annually over ten years as (open circles) calcium ammonium nitrate, (open triangles) urea or (open squares) ammonium sulphate. Closed symbols denote plots also treated with 12.6 tonnes ha^{-1} magnesium limestone in the first year.

incubated soils, a large reduction in the concentrations of magnesium, potassium, sodium, silicon and manganese in soil solution. Liming also stimulated the mineralisation of organic nitrogen as indicated by increased nitrate concentrations. Similar effects were still apparent for pasture soils from a field experiment which was last limed 16 years previously. The influence of land improvements on water quality can be substantial and long lasting. This is especially so for 'sensitive' upland areas (Hornung, Stevens & Reynolds, 1986).

Spatial pH variations

An undisturbed soil profile often follows a general decline in pH with depth with a slight rise again at the profile's base. Mixing upper soil layers through cultivation and addition of lime or fertilisers can cause dramatic changes in soil pH to occur over relatively short distances in a soil profile. Despite the often quoted 'mobile' nature of calcium in soils, when lime is applied to grassland, for example, only a limited depth of penetration occurs. This can cause serious management problems in long lived crops such as orchards where incorporation would cause root damage. Neilsen, Hoque & Drought (1981), for example, noted the slow migration of surface applied lime (6 tonnes $Ca(OH)_2$ ha^{-1}) caused no change in pH beyond 100 mm depth over 5 years.

In many tropical situations, the possibility of strongly acidic subsoils must be taken into account before major benefits from liming are observed. Deep incorporation of lime may well be either extremely difficult technically or too expensive. It is possible that, in these situations, liming the topsoil will only give a limited success because the restricted penetration by roots into the subsoil will markedly increase the likelihood of drought stress.

The spatial (horizontal) variations in pH on a field scale may be substantial. They may be the result of local differences in soil characteristics such as depth, texture or parent material or of poor management practices such as uneven liming or cultivation. It is essential therefore that any soil sample collected should be as representative as possible. There are various sampling procedures designed to minimise these potential sources of error (see, for example, ADAS (1983)).

Soil pH can often vary by two units over very small distances. Neilsen and Hoyt (1982) discussed soil pH variation in British Columbia apple orchards where pH increased with distance from the tree trunk. The variation in soil pH between trees was frequently significant for both widely separated trees in orchards, where pH was expected to be variable, and for adjacent trees in orchards expected to have uniform soil pH. A survey of soils from East Malling by White (1961) showed that an individual field of 0.45 ha had pH values covering the range 4.5–7.9. It was calculated that liming this field to an average of 6.2 would still leave 43% of the area at a pH of 5.7 or less.

Evaluation of soil fertility

Any attempt to evaluate available nutrients requires a sound understanding of the basic soil processes discussed in earlier chapters. The relative importance of an individual process varies not only with time, but also depends considerably upon the element involved and many local soil factors. There are a variety of approaches that may be used to assess how the likely plant requirements match soil nutrient availability. The suitability of either soil or plant chemical analysis varies with factors such as crop type and the specific nutrient involved.

It is important to be quickly aware of nutrient deficiencies. This is particularly the case for short-term, annual crops, where the development of visual symptoms may be too late for any remedial action to be taken. The overall aim of advisory analysis is to ensure that the current crop's nutrient requirements are essentially met through supply from the soil. All methods of soil and plant tissue analysis have their own particular advantages and disadvantages. Foliar analysis is usually preferred for long-lived crops (e.g. many tree and plantation crops), where the often substantial internal nutrient reserves play an important role (Millard and Neilsen, 1989). It is, however, essential to have a well-planned sampling strategy, because concentration is often very dependent upon season and leaf age. As a result, foliar analysis tends to correlate better with previous, rather than the current year's soil data (Atkinson, 1988) but will provide precise information on the plant's overall nutritional health.

In contrast to this, analysis of soil can really only characterise its potential to supply nutrients. A degree of judgement is therefore required to take into account possible growth limiting factors. In all cases analytical results must be assessed against a sound knowledge of local crop/fertiliser responses. This is usually achieved through extrapolation from field-based experimental data acquired over many years. The expense involved in running these experiments usually means some degree of compromise has to be made. In order to improve fertiliser recommendations additional information such as soil type, previous and present cropping and fertiliser histories is required.

Soil fertility

Soil analysis

In spite of the constraints outlined above, there are still two principal reasons for carrying out soil chemical analysis. It may be conducted to exploit a known relationship between soil test results and crop response for predicting fertiliser requirement. Alternatively it is sometimes possible to make a prediction from an analytical result based upon a sound understanding of fundamental chemical and physical principles governing nutrient availability (Olsen and Khasawneh, 1980).

It is useful to subdivide the soil nutrient pool into a number of categories based upon the ease with which the individual elements can be extracted. While the resulting division is somewhat artificial, it does provide a basis for a useful relationship to be made with potential plant uptake. Three subdivisions are common:

$$\text{Soil solution X} \underset{}{\overset{K_1}{\rightleftharpoons}} \text{Labile soil X} \underset{}{\overset{K_2}{\rightleftharpoons}} \text{Non-labile soil X}$$

where X represents an individual element, K_1 and K_2 are equilibrium constants, and labile implies being able to show a rapid exchange with isotopically labelled, but otherwise identical, ions. Any particular soil will strive to attain a state of equilibrium. Additions to (fertiliser, atmospheric deposition) and removal from (plant uptake, leaching, exchange/sorption reactions) the soil system ensure an equilibrium situation is rarely, if ever, achievable.

The immediate nutrient requirements of a plant are primarily met through supply from the soil solution. The total amount of a particular element in the soil solution varies considerably but for many elements such as phosphorus it is small and therefore can be easily depleted if the rate of renewal from the labile pool (soil surface exchange sites) is insufficient. It is essential then that the rate of replenishment matches the rate of depletion. In order to give some idea of the required turnover rates, it is worth considering a simple example which involves the phosphorus dynamics for a grass/clover field experiment. The annual removal of phosphorus in three cuts was approximately 17 kg ha^{-1}, representing a daily uptake rate equivalent to 85 g ha^{-1}. Table 5.13 shows that the non-labile phosphorus pool is three orders of magnitude greater than the labile phosphorus pool, and less than 1% of the latter is present in soil solution at any one time.

The continued supply of nutrients to a root is achieved through a combination of two mechanisms, which both rely on gradients created by an actively absorbing root surface. Mass flow of water and the dissolved solutes it contains will occur as a direct result of water movement through the soil to meet transpirational needs, or as a consequence of any other hydraulic gradient. This will be especially important for ions that exhibit relatively high concentrations in the soil solution (e.g. nitrate) and are therefore likely to show only a small concentration depression at the root surface (Mengel, 1982).

The second mechanism, diffusion, occurs as a consequence of localised depletion and therefore lower concentrations around the root (e.g. potassium and phosphorus). This will stimulate desorption of ions from the soil particulate surface. It is also possible that certain elements may show a marked build up in concentration (Marschner, 1991). An extreme example is the formation of calcium sulphate/carbonate pedotubles around roots (Barber, 1984). The zone of depletion (or accumulation) is very dependent upon the ion involved, soil and crop type and time (Walker & Barber, 1961; Farr, Vaidynathan & Nye, 1969). The diffusion rate of ions in soil solution is dependent upon factors such as moisture content, tortuosity of flow pathway, and degree of interaction between the ion in question and the soil surface exchange complex (Nye & Tinker, 1977). Typical diffusion coefficients for NO_3^-, K^+ and $H_2PO_4^-$ are 10^{-6}–10^{-7}, 10^{-7}–10^{-8} and 10^{-8}–10^{-11} $cm^2 s^{-1}$ respectively.

It is readily apparent that soil and solution in close proximity to actively growing roots will experience a very different set of conditions from those of the bulk soil. Differences in the uptake balance between cations and anions can directly influence local soil pH (Moorby, White & Nye, 1988). Rhizosphere pH may differ at the root surface to a few millimetres away by 1–2 units as a result of H^+ or HCO_3^- efflux (Nye, 1981).

The ease with which a soil can maintain a near steady state concentration in response to depletion from the soil solution has been termed its buffering capacity. The term is represented by:

$$\text{Buffer capacity} = X_1/X_2 = (Q)/(I)$$

Table 5.13. *Typical phosphorus pool under grass/clover*

	Non-labile P	Labile P	Soil solution P
	40 000 kg ha^{-1}	40 kg ha^{-1}	62.5 g ha^{-1}
Turnover time		3 years	1 day

where X_1 is the exchangeable fraction, i.e. the quantity factor (Q), and X_2 is the equilibrium soil solution concentration, i.e. the intensity factor (I). The buffering capacity is dependent upon a number of soil characteristics, the relative importance of which varies with individual elements. The best method of establishing this relationship is by using plant uptake over time in pot or field experiments, but this can be expensive and time consuming. Other, purely chemical approaches are possible but can suffer from not being able to simulate time aspects adequately. In particular ion exchange resins offer a useful approach (Barrow & Shaw, 1977; Somasiri & Edwards, 1992). The resin, acting as a sink for exchanging ions, has been suggested to mimic the pattern of uptake by roots (Amer, Bouldin, Black & Duke, 1955).

Soil testing has tended to be independently organised in many countries and laboratories, resulting in the development of a wide variety of extraction procedures, often with only a local application (Cottenie, 1980). This sometimes makes result comparison and interpretation of results between research groups and countries difficult. An ideal extractant for use in assessing soil nutrient status should be capable of relating to some definable fraction (in terms of plant uptake) of the particular element in question. In practice, however, this is seldom achievable for all situations.

The frequency of soil sampling depends upon crop type and nutrient demands. For example the UK Agricultural Development and Advisory Service (ADAS, 1983) suggests every 7 years for permanent pasture, every 3–4 years for intensive grass and arable rotation, down to 2–3 years for intensive outdoor horticulture. It is important to collect samples before the most demanding and valuable crop. The majority of soil samples in Britain are collected for analysis during the early autumn. During the early 1980s, 70% of the soil analysis undertaken annually (England and Wales) was carried out by fertiliser firms; ADAS only accounted for 6% of the total (Church, 1981). The value of any soil sampling is greatly dependent upon its representativeness. Soil heterogeneity is partly overcome by the soil fertility rating systems, which tend to allocate soils to broad bands, each one covering a wide range of extraction values (Cope & Evans, 1985). It is generally accepted that laboratory precision, and hence error from analysis, is much less of a problem than the uncertainty associated with field sampling and soil heterogeneity (Cipra, Bidwell, Whitney & Feyerherm, 1972).

For trace elements, soil analysis is often conducted where deficiency or toxicity is suspected. The extractants used are the result of painstaking pot or field experiments using many soils. The aim of such trials is to find an extractant that removes from soils amounts of the element of interest which correlate well with its uptake by plants growing on the same soils.

Thus hopefully the extractants take into account the quantity factor to an appropriate degree, and provide an indication of the amount of element that will become available during growth. For example, ethylenediamine-tetraacetic acid (EDTA) solution is widely used for the estimation of plant-available copper, zinc, iron and manganese. Some authors however prefer diethylenetriaminepentaacetic acid (DTPA) for copper and zinc. Hot water is most commonly used to estimate available boron. Available molybdenum is extracted with acid ammonium oxalate. A soil analysis text, such as the excellent two volumes on *Methods of Soil Analysis* produced under the auspices of the American Society of Agronomy, should be consulted for further details on these and other soil analyses.

6

Soil chemistry and freshwater quality

One of the many vital roles played by soil is its function as a buffer regulating the quality of water in rivers and lakes and, to some extent, groundwaters. In the majority of catchments, unless outcropping rock is dominant, precipitation interacts to a substantial degree with soil. The precise fate of incoming precipitation, i.e. the pathway it follows through or over soil to a drain, stream or lake, depends upon the physical characteristics of the soil, the quality, duration and intensity of precipitation, prior climatic conditions, surface and subsurface topography, and the physical form of the precipitation (e.g. rain, hail, snow, mist etc.). The hydrological pathway followed, in turn, governs which zone or zones of the soil the water interacts with en route to open surface water or groundwater. The purpose of this chapter is to consider briefly how the interactions between draining water and the soil biota, atmosphere, minerals and organic matter ultimately influence water quality in natural ecosystems. We will then be in a better position to see how these natural processes cope with human activities such as modern agricultural and forestry practices and with pollution generally. The way in which the soil/water system copes with pollutants of other origins is discussed in Chapter 7.

Drinking water must be acceptable with respect to taste, colour, microorganisms, salinity, acidity and concentrations of organic pollutants, toxic heavy metals, aluminium, nitrate and nitrite. Clearly some of these factors are interactive, especially if considered in the light of domestic water supplies. Excessively acidic waters tend to mobilise aluminium either from acid soils or from river beds, for example. They may also contain high concentrations of lead from lead plumbing fittings. Even very low concentrations of some organic pollutants may impose an undesirable taint, a problem which may sometimes be exacerbated by treatment processes such as chlorination.

Aluminium in drinking water is now recognised as a potentially serious health hazard, although it has previously always been regarded as non-toxic and not absorbed from the gastrointestinal tract (Elinder, 1984). Accumulation of the element in the body, especially in the brain, may adversely affect the central nervous system, leading to symptoms such as dementia, aphasia, ataxia, and convulsions. A number of instances have been recorded where patients with kidney disorders have been seriously adversely affected by aluminium in dialysis water (Elinder, 1984).

Unlike aluminium, the toxic heavy metals such as cadmium, mercury and lead are generally present at low concentrations in most unpolluted soils, and even lower concentrations in freshwaters. Any changes in soil, whether natural or as a consequence of human activities, which lead to increased mobilisation of the toxic heavy metals into drinking water supplies must be considered as a potential health risk. A useful concise account of the effect of these metals on human health has been published by Elinder (1984).

Of the inorganic nitrogen-containing species found in water supplies, nitrate is generally thought to be the least toxic and nitrite the most potentially dangerous. The maximum admissible concentration (MAC) values for ammonia (including ammonium), nitrate and nitrite are 0.5, 50 and 0.1 mg l^{-1} respectively (Water Act, 1989).

The greater part of the water supplied to domestic consumers is not, in fact, used for drinking purposes, but rather for washing, waste disposal and garden watering. The latter uses impose lower quality control requirements, but this is of academic interest since in most urbanised areas houses are traditionally only linked to a single water supply.

In many respects, water quality requirements of freshwater fish are not dramatically different from those for drinking water for human consumption, although moderate toxicity symptoms are more likely to pass undetected. Severe toxicity may result in loss of entire populations or age groups in populations. Such observations helped to trigger the research thrust into the causes of freshwater acidification in the 70s and 80s (Cresser & Edwards, 1987; Mason, 1991). As with human health, mobilisation of soil aluminium into waters is now recognized as a potentially serious problem, although for fish the extent of such a problem depends on the water pH and ambient calcium concentration (Harvey, 1985).

The hydrological cycle

On a global scale, water is continually being cycled and recycled. The relevant processes should already be clear from the discussions of element cycling in Chapter 4. Evaporation from the oceans, lakes and rivers, and soil and wet rock surfaces, and evapotranspiration from plants lead to transfer to the atmosphere. There some water is transported in the vapour phase, some, after condensation to form clouds, in the liquid phase. Condensed-phase water may return to the soil, vegetation or open water surfaces as precipitation in various forms, including rain, sleet, snow and intercepted mist. Vapour-phase water may return as dew or frost. The relative amounts of precipitation and evapotranspiration play an important role in soil formation. Where the former is greater, there is a natural tendency in the long term towards soil acidification as base cations are leached out of the ecosystem in drainage waters. These pass to rivers, lakes or groundwaters. Water in the majority of rivers eventually returns to the oceans, apart from that which is lost by evaporation. When evapotranspiration exceeds precipitation, there is a natural tendency towards the development of alkaline soils with calcium carbonate accumulation (see Chapter 4).

Factors affecting hydrological pathways

In the simplest possible situation, rain falling onto the flat surface of a freely draining soil will first wet up the surface of the soil. If the rain is light and the soil dry, the wetting front will move downwards without necessarily saturating the soil. As a soil dries out, the residual water is retained progressively more and more tightly in smaller and smaller soil pores. It may, however, be removed by applying suction slightly greater than the force retaining the water, the so-called soil water tension. As the soil wets up, the soil water tension falls. Provided the pore network is more or less continuous, the tensional forces tend to equilibrate, thus contributing to the redistribution of the water and the downwards movement of the wetness front. Thus eventually the water would drain down to the water table.

If the rain is heavy and prolonged, the soil at the surface eventually may become saturated, with all the soil pores completely filled with water. Once this stage is reached, further precipitation can only drain completely down through the soil if the infiltration rate equals or exceeds the rate at which the precipitation is falling on the surface. Infiltration rate under saturation conditions is expressed in terms of a parameter known as the saturated hydraulic conductivity. If the saturated hydraulic conductivity is too low, puddles form at the surface. If the surface is sloping, rather than horizontal, water flows laterally downslope over the soil surface. Such flow is known as overland flow or surface runoff. The flow velocity of surface runoff at an upland site depends upon the vegetative cover. For moorland vegetation it is easy to envisage that the flow velocity will be appreciably lower than that of open channel flow on the same slope, as the water flow path twists and turns around the plant stems. Thus there is usually plenty of time for surface runoff to equilibrate with the surface soil, since cation-exchange reactions (see Chapter 4) of soils take place quite rapidly.

The saturated zone during a storm event does not necessarily occur at the soil surface. It may occur below the surface at any depth at which the vertical hydraulic conductivity suddenly falls by an appreciable amount. The point at which the soil meets impermeable bedrock may be regarded simply as an extreme example of such a situation. When it occurs, lateral, subsurface flow results. If the arresting boundary is not bedrock, perched water tables may be formed. These conditions may arise when relatively permeable organic horizons overlie less permeable mineral or organic soils, or at cemented or indurated horizons. An indurated horizon is a layer of soil in which the soil is packed to give a high bulk density, and thus often a low hydraulic conductivity, but the individual mineral grains are not necessarily chemically cemented together.

When perched water tables do occur, water at the bottom of extended slopes may be subject to considerable water pressure because of the hydraulic head. The pressure may be sufficient to cause throughflow to break back through the surface, at which point it becomes overland flow. Although such water rapidly reequilibrates chemically with the surface soil, its solute composition may differ from that of simple surface runoff because of the species dissolved in the underlying soil horizons. The major hydrological pathways discussed above are represented schematically in Fig. 6.1.

Rainstorms, snowmelt and freshwater quality

When a prolonged and heavy rainstorm occurs in a flat region with freely drained soils, the hydrological pathway is unlikely to change significantly unless flooding eventually occurs. The water will drain to the water table, flushing out soluble salts previously present in the soil solution. The soluble salts composition will be the net result of previous wet and dry deposition inputs, geochemical weathering, microbial activity, plant uptake, the presence of any soluble fertiliser, climate prior to the precipitation event, and the precipitation composition. Initially therefore there may be a small increase in soluble salts in riverwater through the flushing effect, but the extent of this will depend upon the degree of mixing with groundwater and catchment size.

In upland catchments in regions with humid climates, substantial changes to hydrological pathways are much more probable in response to storm events. In light rain and between storms, water tends to drain to underlying mineral soils, as for flatter sites. En route, the soluble salts concentration tends to increase as a consequence of geochemical weathering. The calcium, magnesium, potassium and sodium bicarbonates formed by weathering, and the base cation inputs from the atmosphere and microbial litter decomposition equilibrate with the adjacent soil cation-exchange complex. Some potassium may be fixed by secondary clay minerals. Soluble silicate increases to an equilibrium concentration. The drainage water becomes highly enriched in carbon dioxide as a consequence of the high concentration of the gas (1–2%) often found in the soil atmosphere.

In heavy rain storms in upland catchments, changes in hydrological pathways are very much favoured by the steep slopes, frequent outcrop-

Fig. 6.1 Important hydrological flowpaths in an upland catchment: (1) overland flow; (2) return flow; (3) through flow; (4) percolation to ground water.

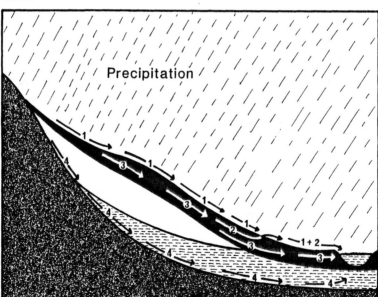

ping rock with negligible water retention capacity, thin soils on upper, steeper slopes, and generally higher precipitation and lower evapotranspiration. Thus as the storm proceeds and river discharge or lake level rises, often within an hour or two for smaller catchments, the contribution of throughflow from more acidic and organic rich surface horizons increases dramatically. The water chemistry increasingly becomes dominated by equilibria with organic rather than mineral material. Thus, as discharge rises (Fig. 6.2), the concentrations of

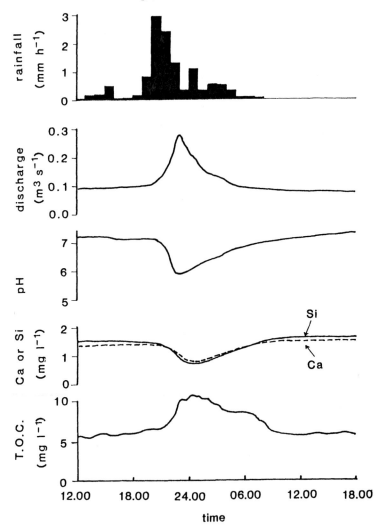

Fig. 6.2 Typical hydrograph and riverwater chemical composition changes in response to a storm in an upland catchment in northern Europe or America. TOC denotes the total organic carbon in solution and fine suspension form.

dissolved organic matter and the elements associated with it *via* organic matter complexation, especially iron and aluminium, increase. Elements such as silicon and calcium, whose concentrations primarily depend upon geochemical weathering, become progressively more dilute. The water becomes increasingly acidic. When the storm ends, discharge subsides and all the changes described above are reversed (Cresser & Edwards, 1987). One study in northern Sweden has shown that quite small zones of acid organic soil adjacent to river banks may contribute very significantly to acid flushes in rivers in heavy storms (Bishop, Grip & O'Neill, 1990).

The same sequence of events described above occurs in response to rapid snowmelt in many upland catchments. The meltwater equilibrates with the surface soil. Because the first meltwaters are often from snow with the highest soluble salt concentration (the snow with the greatest depression of freezing point), the concentration of hydrogen ions displaced from the cation-exchange sites is especially high. Moreover, additional salts, thought to be of ruptured cell origin, may be solubilised as a consequence of the soil freeze–thaw process itself (Edwards, Creasey & Cresser, 1986). If the soil were frozen to depth before the snowfall initially occurred, frozen subsoil might contribute to the occurrence of a perched water table very close to the surface. It might thus facilitate the generation of acidic drainage waters from surface horizons. It must be remembered that soil under snow is not always frozen, however, even after prolonged periods with very low air temperatures, because of the insulating effect of the snow cover.

Buffering of pH in upland streams

An important consequence of the pathway changes during episodes described above is that upland rivers are generally acidic only for a number of relatively short periods each year. The net effect of this may be most clearly seen if results of regular sampling over a year are presented as pH duration curves, as in Fig. 6.3. This shows, for each of three streams in northeast Scotland, for what percentage of the total time the water pH was below any specified value.

One aspect of water chemistry that often surprises soil chemists when they meet it for the first time is the observation that water draining from a mineral soil with a pH of say 5.5 may, in the resulting river, have a pH in excess of 7. The pH rise is a consequence of the elevated concentration of carbon dioxide in the soil atmosphere. Carbon dioxide dissolves in water to produce carbonic acid:

Buffering of pH in upland streams

$$CO_2 + H_2O \rightleftharpoons H_2CO_3 \quad (6.1)$$

$$K_1 = \{CO_2\}\{H_2O\}/\{H_2CO_3\} = \{CO_2\}/\{H_2CO_3\}$$
$$\text{(since } \{H_2O\} = 1\text{)} \quad (6.2)$$

Carbonic acid then dissociates:

$$H_2CO_3 \rightleftharpoons H^+ + HCO_3^- \quad (6.3)$$

$$K_2 = \{H^+\}\{HCO_3^-\}/\{H_2CO_3\} = \{H^+\}\{HCO_3^-\}K_1/\{CO_2\} \quad (6.4)$$

Suppose that the carbon dioxide in the soil atmosphere, $\{CO_2\}$, is enriched 100 fold compared with the atmosphere above ground. When the water drains out of the soil, $\{CO_2\}$ falls 100 fold. Since K_1 and K_2 are constant, from equation (6.4) the product of $\{H^+\}$ and $\{HCO_3^-\}$ must also fall 100-fold. If this happens simply by equations (6.3) and (6.1)

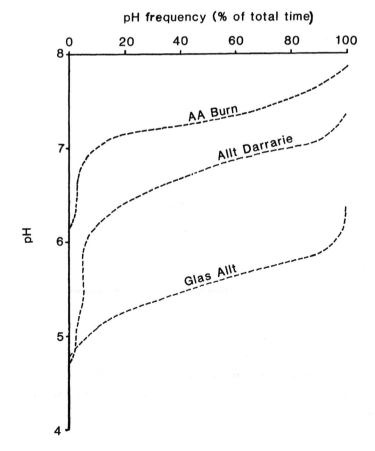

Fig. 6.3 The pH duration curves of three upland rivers in northeast Scotland, showing the relative importance of short duration acid episodes.

moving from right to left, both $\{H^+\}$ and $\{HCO_3^-\}$ will fall 10 fold. Thus the pH will rise by one unit. If, on the other hand, the soil solution contains bicarbonates from geochemical weathering, $\{HCO_3^-\}$ will rise by less than 10 fold, and $\{H^+\}$ by more. The pH will then increase by more than one unit. The effect may easily be demonstrated, in practice, by equilibrating river water with atmospheres containing known carbon dioxide concentrations and monitoring the pH change (Cresser & Edwards, 1987). When the equilibrating atmosphere carbon dioxide concentration matches that of the soil atmosphere, the water pH should match the soil pH.

Figure 6.3 is a good example of why it is important to understand soil–water chemical interactions. Once the acidification mechanism is understood, it becomes obvious that acidification events will occur in poor weather conditions, either during heavy storms or rapid snowmelt. Sampling rivers only on sunny, summer days would give no indication of the acid episodes likely to occur under different climatic conditions. Also important, from the viewpoint of fish status, is that acid conditions are most probable under conditions where input of calcium from mineral weathering is likely to be minimal, and output of organically complexed aluminium is likely to be high.

Catchment characteristics influencing soil–water interactions

Although we have now considered how soils influence the solute composition of drainage waters in general terms, it is appropriate at this point to look at the contribution of individual catchment characteristics to the overall process. These may be associated with precipitation characteristics, such as amount, distribution with time, chemical composition or proportion of snow, with other climatic factors, such as temperature, with soil characteristics, such as drainage, thickness or chemistry, with general catchment characteristics, such as slopes, aspect and topography, or with vegetation characteristics. Each of these aspects will be considered briefly in turn.

Precipitation characteristics

From the preceding sections, it should already be apparent that catchments with high annual precipitation are more likely to give rise to acid waters deficient in calcium and high in iron and aluminium than catchments in drier areas, other factors such as soil type and drainage basin topography being equal. This is so because of the greater probability of surface runoff and return flow occurring (see Fig. 6.1). It should also be remembered that, as a general rule, the evolution of acid soils from the same parent material over the same timescale will be more advanced in an area with higher annual precipitation. If more water has contacted the mineral soils, more mineral weathering will have occurred, and more base cations leached from the ecosystem. Also physical erosion is more likely in areas with higher annual precipitation, leading to thinner mineral soils on upper slopes. The greater degree of soil evolution may manifest itself as more clear-cut podzolisation or by greater incidence of deep peat coverage of the mineral soils. The greater the contribution to total discharge from water draining from or over peat, the greater the dissolved organic matter content of the water.

The above comments obviously only apply to catchments in which precipitation exceeds evapotranspiration. When the reverse situation is true, as we saw in Chapter 4, alkaline soils containing free calcium carbonate are likely to develop. Thus drainage waters in such areas, when drainage occurs, are likely to contain quite high concentrations of calcium and, to a lesser extent, magnesium and other base cations. Substantial natural organic matter accumulation is not likely in arid climates, so organic matter concentration in solution is likely to be low. A notable exception occurs when sodic soils are irrigated and drained. The very high pH causes substantial dissociation of phenolic and carboxyl groups on the organic matter, rendering it much more soluble.

The distribution of precipitation is also important. Acidification episodes and associated problems are more probable when a large amount of rain falls over a few hours than when the same amount of rain falls over a much longer period. The amount of precipitation falling as snow is also crucial, as is the speed of snowmelt. While the snow is lying on the surface, the base cation concentrations in the riverwater are likely to be relatively high, since most of the water may originate under these conditions from deeper, mineral soils. Discharge generally remains low, and water quality is not dissimilar to that associated with summer baseflow, but with one notable exception: nitrate concentrations in winter tend to be appreciably higher in streams draining unfertilised upland catchments than those in summer. In these low nitrogen soils, virtually all of the nitrate input in the precipitation is taken up by plants or by microbial biomass over the active growth season, whereas in winter nitrate tends to pass through the ecosystem.

The problems associated with rapid snowmelt are analogous to those associated with very heavy storm events. Thus waters tend to exhibit increases in concentrations of dissolved organic matter, iron and aluminium and decreases for calcium, magnesium and silicate.

Both in storms and during rapid snowmelt, the concentration of sulphate in river water may increase considerably compared with that during baseflow, especially in areas where the sulphate concentration of the precipitation is high. Acid upland soils tend to contain a so-called B horizon which contains substantial quantities of iron and aluminium hydrous oxides, and has a characteristic rust colouration. As discussed in the section on pH effects on sulphur availability in Chapter 4, these hydrous oxides have a considerable sulphate adsorption capacity. Thus baseflow water which has drained through a mineral B horizon may be depleted in sulphate compared with water draining laterally through a surface organic horizon. We shall return to the topic of sulphate adsorption again in Chapter 7 when discussing the capacity of soils to cope with atmospheric pollution.

In agricultural drainage basins in which the soils are regularly fertilised and limed, the changes which occur in precipitation chemistry from storm to storm, or even within individual storms, (Cresser & Edwards, 1987) are of little consequence to water quality. This is not the case in unfertilised upland catchments in wet climates, however. We have already seen that precipitation nitrate may pass through such a catchment in winter, for example. Mineral weathering may play so little a part in generation of soluble salts that variability in the solute loading of the precipitation becomes significant. This may be the case for deep peats, for example. Under these circumstances, an increase in the loading of sodium chloride in precipitation arising as a consequence of severe offshore storms could lead to more hydrogen, iron and aluminium ions being displaced from exchange sites, and transferred to riverwater in lateral-flowing drainage water. Thus a sudden increase in sodium chloride in rain could lead to a particularly severe acid episode (Wright, Norton, Brakke & Frogner, 1988). Note that the effect is due to competitive ion exchange, which was discussed in Chapter 4.

Other climatic factors

Temperature obviously plays an important role in regulating soil–water interactions. The higher the temperature, the greater the evapotranspiration and the less the probability of surface runoff, return flow or throughflow occurring. Biological activity also increases at higher temperatures, so soil formation processes will be modified. The most conspicuous effect on water quality in natural ecosystems is perhaps that on riverwater nitrate concentrations. A typical example is shown in Fig. 6.4, which illustrates the seasonal effect on monthly nitrate output from a moorland catchment in northeast Scotland. During the active growth period between May and September, very little nitrate leaves the catchment, although inputs in rain are considerable, except in catchments grossly polluted with nitrogen over many years. The effect of temperature on biological activity means that peat accumulation is likely to be slower in a warmer region. Snow accumulation may be much less, so problems associated with rapid snow melt may be avoided unless an area has long, very cold winters and a very rapid temperature rise in spring. High summer temperatures may lead to the temporary drying out of boggy areas, with subsequent very high levels of microbial activity when they wet up again, which may mobilise substantial amounts of acids, e.g.:

$$RSH + H_2O + 2O_2 \rightleftharpoons ROH + SO_4^{2-} + 2H^+$$

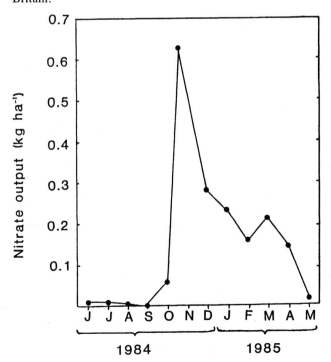

Fig. 6.4 An example of the distinctive seasonal trend in monthly nitrate outputs in riverwater from a typical upland moorland catchment of northern Britain.

$$RNH_2 + 2O_2 \rightleftharpoons ROH + NO_3^- + H^+$$

$$4Fe^{2+} + O_2 + 10H_2O \rightleftharpoons 4Fe(OH)_3 + 8H^+$$

If the dried-out boggy area floods and is low-lying, substantial quantities of acid may be transported into adjacent streams or lakes.

Another important climatic effect on soils of relevance to water quality is the impact of natural freeze–thaw cycles (Edwards *et al.*, 1986), which was mentioned briefly earlier in this chapter. These cycles may mobilise significant quantities of nitrate, organic matter, iron, aluminium and organic carbon.

Soil characteristics

Both chemical and physical characteristics of soils influence the solute composition of the associated drainage waters. It is appropriate to consider both together here since they are interactive to a large degree. For example, soil drainage characteristics are regulated by physical restrictions, but have far-reaching effects upon soil chemical properties. Other aspects of soil water movement are also generally regarded as falling within the remit of the soil physicist rather than that of the soil chemist, but they too influence soil formation processes.

The most important physical parameter in the present context is hydraulic conductivity, and its change with depth down the soil profile. This, together with slope, controls the hydrological pathway under any specified set of climatic conditions. It therefore controls, in the short term, what soil(s) the drainage waters equilibrate with. In the longer term, it has a major impact upon the chemical properties of those soils at any stage in their evolution (see the sections on element cycling towards the end of Chapter 4).

Soil thickness above rock may be regarded as an extreme case of depth to a sharp change in hydraulic conductivity, assuming the rock is of low porosity. Deep, freely drained mineral soils in humid regions are likely to produce near neutral drainage waters by mineral weathering equilibria for many thousands of years. Depending upon the rock subsurface topography and watertightness, water may drain out of a catchment or may simply add to the water at the water table. When the latter process occurs, the presence of elevated nitrate concentrations in the drainage as a consequence of agricultural or forestry management practices may be a cause for concern if the nitrate starts to accumulate and the groundwater is used for domestic supplies. Unless the mineral soil is very old (i.e. on a geological timescale), or shallow, or the soil was formed from highly weathered, acid rocks or associated tills, the base status and pH of such waters is likely to be high, and the dissolved aluminium concentration very low.

Deep, mineral soils are less likely on steep slopes because of the greater probability of physical erosion occurring. Thin mineral soils have a limited capacity for storing water during prolonged heavy precipitation events, even if they are freely draining. Thus the risk of surface runoff, throughflow or return flow occurring is much greater on steep slopes in wet regions. This is also true where the depth to a horizon of low hydraulic conductivity is limited.

Soil parent material

It should be clear from the discussion of changes in soil pH in Chapter 4 that, for unfertilised soils, soil pH depends upon the capacity of soil mineral geochemical weathering to replace base cations depleted in drainage water or by erosion or crop or animal removal. It might be expected then that soil base saturation will depend upon the current state of soil evolution, which depends, other formation factors being equal, upon the nature of the soil parent material. This is indeed the case. Figure 6.5 compares the seasonal trends in river water acidity for two catchments of similar age, topography, climate and land use.

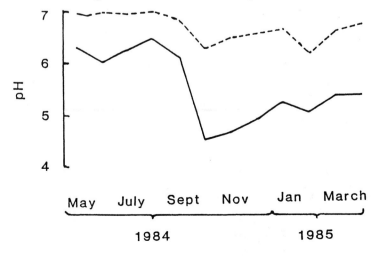

Fig. 6.5 The influence of soil parent material on riverwater pH over a particularly wet autumn. The solid line shows data from a catchment with soils derived from granite, whereas the broken line shows data for a river draining soils from a more base-rich quartz-biotite-norite.

However, in one drainage basin (Glendye) the soils have evolved from the acid rock granite and associated tills. In the other (Peatfold) they are derived primarily from a much more basic quartz–biotite–norite. At Glendye, podzolisation is more advanced, as is the evolution of acid hill peats. The effect that this has upon drainage water chemistry is readily apparent in Fig. 6.5. The greater extent of peat evolution also leads to more frequent occurrence of water at Glendye with elevated concentrations of dissolved organic matter and organically complexed iron and aluminium.

A useful indication of the impact of soil parent material upon water quality in British rivers may be obtained from the concise but comprehensive review by Walling & Webb (1981). These authors quote annual outputs (t km^{-2} yr^{-1}) from 47 rivers of calcium, magnesium, sodium and potassium. For calcium, which usually is the most mobile element relative to the amount in parent material, the results varied between 0.5 and 100 t km^{-2} yr^{-1}. This suggests lime requirements of between 12 and 2500 kg ha^{-1} for the catchments to maintain soil calcium status. Water draining soils over granites and acid metamorphic rocks, and sandstones, slates and shales invariably exhibited low conductivity values (Walling & Webb, 1981).

Although not a parent material in the strictest sense, lime and its probable effects upon water quality in upland areas in humid regions may be appropriately considered at this point. When lime is applied to upland acid peats without cultivation, its effect is generally confined to the surface horizon because the calcium is so tightly bound to the organic matter (Sanger, Billett & Cresser, 1992). This may have a very beneficial effect upon water quality in catchments where surface flow processes are important. Water pH and calcium concentration rise considerably. Increased mineralisation of organic nitrogen may lead to nitrate mobilisation, however. Management practices leading to reduction of the use of lime in upland areas may conversely lead to marked deterioration in freshwater quality in upland lakes and rivers. Crawshaw (1986) has discussed the possible significance to fish populations in the River Esk and its tributaries of the removal of the liming subsidy in Britain.

Catchment characteristics

The important aspects of drainage basins in the present context, slope and other surface and subsurface characteristics, aspect and altitude, are inextricably linked to factors which have already been discussed. Altitude, for example, influences precipitation type, amount and distribution, temperature and vegetation growth and hence water and nutrient element use. These factors, in turn, influence soil thickness and other soil formation factors. Slope influences the probability of physical erosion, and hence soil thickness. It thus both directly and indirectly influences water retention in the soil and hydrological pathway. Models such as TOPMODEL have been developed to allow quantitative prediction of the link between river catchment topographic characteristics and water chemical parameters (Wolock, Hornberger & Musgrove, 1990).

Modelling water quality

From the discussion on the previous few pages it appears that attempts to model streamwater solute chemical composition need to take into account both the chemical properties of soils in a catchment and the catchment characteristics regulating hydrological pathways; this is indeed the case in practice. Consider Fig. 6.6, a soils map for the Allt a'Mhaide drainage basin, a typical small upland catchment in northeastern Scotland (Billett & Cresser, 1992). Water entering the

Fig. 6.6 Soils map for a typical small upland catchment in the UK.

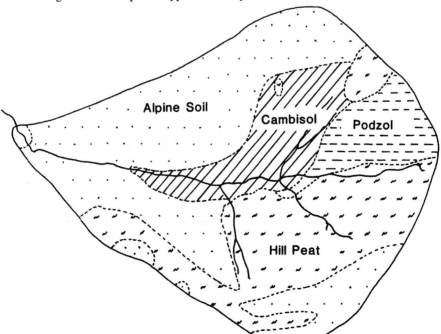

stream has originated within at least four very different soil types: hill peats, podzols and alpine soils which are all very acidic and more base-rich cambisols. Each soil type has its own distinctive characteristics.

Figure 6.6 also shows the importance of soil variability and processes of soil formation. The large patch of cambisol in this catchment reflects the receiving nature of this soil volume. Mobilisation of base cations from alpine soils or podzols on upper slopes as a consequence of natural leaching or of acid deposition may, for many years, maintain or even increase the base status of soils such as cambisols or gleysols on lower slopes. Moreover, because these receiving sites are often adjacent to streams, they may have a disproportionate effect (compared to their relative area) upon surface water solute composition.

Billett & Cresser (1992) took this variability into account in the modelling approach that they adopted. They applied a two-compartment model to each main soil type in each of the ten catchments that they studied. They then generated 100 points randomly over each catchment and estimated the flow path from each point to the river or one of its tributaries (Fig. 6.7). To each point they ascribed an appropriate proportion of the total amount of water entering the catchment, and

Fig. 6.7 Predicted flowpaths from random points in the same catchment mapped in Fig. 6.6.

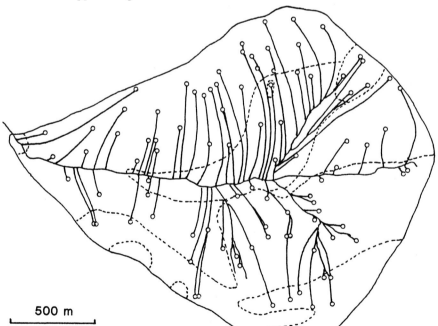

they assumed that the soil (two compartments) at the bottom of the flowpath, immediately adjacent to the stream, regulates the chemistry of that portion of water before it enters the stream. They further assumed that the solubility of each cation is regulated by its relative contribution to the total exchangeable cation population; thus calcium solubility is regulated by the ratio of exchangeable calcium to cation exchange capacity, etc. The 100 such values were then volume weighted, to provide, for example, a weighted value of exchangeable calcium as a percentage of CEC, (i.e. weighted % Ca saturation). The approach was applied to examine the relationship between minimum and mean concentrations of base cations or H^+ and the appropriate percentage saturation parameters, using for calibration purposes weekly river quality data over two years and detailed soil survey/ analysis data obtained from the ten catchments. Figure 6.8 shows a typical calibration graph for the model, which could then be used to make predictions for other rivers in the region using only catchment soil data as an input.

The approach outlined, which is known as the Aberdeen Soil Horizon Model, is one of several models available for prediction of riverwater composition. It is limited to relatively local regions by the lack of incorporation of a mobile anion effect. Precipitation events with elevated strong mineral acid anion concentrations would result in more

Fig. 6.8 Relationship between maximum riverwater calcium concentration observed for each of ten catchments in northeast Scotland over two years and weighted soil calcium saturation (see text for explanation).

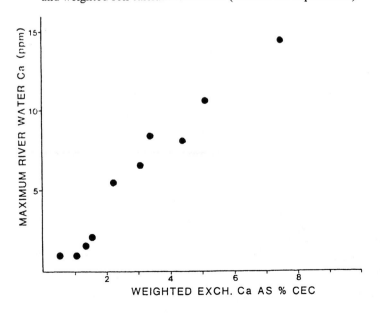

cations, including H^+ ions, being displaced from soil cation-exchange sites into drainage water, as mentioned earlier.

Where computer models have been used to investigate acid deposition effects upon upland streams, mobile anion concentration is invariably regarded as one of the key factors regulating water pH. The Birkenes Model, which evolved over 15 years primarily from data for a small drainage basin in southern Norway, was one of the earlier successful models of water solute chemistry (Stone *et al.*, 1990). Conceptually, it is extremely simple, the catchment being regarded as consisting of two soil reservoirs, an upper one providing rapid flow (surface throughflow/overland flow) and a lower reservoir providing baseflow water. Snow is treated as a third reservoir where appropriate. Thus the model predicts water quality by consideration of mixing from two or three compartments. It takes into account evapotranspiration and piston flow (positive water displacement from the lower compartment by a hydraulic head in the upper one).

While the Birkenes model, once adequately calibrated, is very valuable for prediction of short-term changes in water quality with time, longer-term predictions of acid deposition effects upon water quality require a reliable quantitative estimation of changes in soil chemical properties, especially base saturation and exchange acidity, over time. Recently a model called SMART (Simulation Model for Acidification's Regional Trends) has been described for the elucidation of temporal and geographical patterns of forest soil acidification in Europe (de Vries, Posch & Kämäri, 1989). SMART is a simple process-based dynamic model with modest data input requirements. It depends upon a series of mass balance equations which describe input–output relationships for cations and strong acid anions, after making a series of simplifying assumptions. It differs from the Birkenes model in that it is aimed primarily at elucidation of long-term trends (over several years). It incorporates most major relevant processes, including ion deposition, uptake immobilisation, nitrification, weathering, dissolution and ion exchange.

One of the great difficulties with models which predict long-term change (over decades) in soil or river water quality is validation. To overcome this problem, when modelling pollution effects comparisons are sometimes made of catchments in polluted and relatively pristine areas. For example, a comparison has been made of headwater streams at the much-studied Hubbard Brook Experimental Forest in New Hampshire and at Jamieson Creek in British Columbia (Driscoll, Johnson, Likens & Feller, 1988). Both sites are underlain by granitic bedrock, have been recently glaciated and contain thin, acidic soils. Both watersheds have a heavy winter snowpack and are forested. However, there are differences in forest type and precipitation amount. Selection of matched catchments, in which all important characteristics other than pollutant load are identical, has always proved virtually impossible. Nevertheless this study demonstrated a shift in the more polluted catchments to strong acid-regulated acidity and increased mobilisation of toxic inorganic aluminium species. The Birkenes catchment has now been studied for more than 20 years, and a steady decline in calcium concentration over 15 years in a single river is perhaps more readily acceptable evidence for change (Christophersen *et al.*, 1990b).

Christophersen and colleagues (1990a) observed that, although models are frequently used to predict soil water quality, the predictions are rarely verified using field observations. They concluded that where direct measurements of soil water composition were available in Birkenes, Norway, and Plynlimon in mid-Wales, they were insufficient to explain streamwater chemistry.

Vegetation effects

The type of vegetation growing in a catchment may significantly influence water quality in a natural ecosystem (Johnson *et al.*, 1991; Lelong, Dupraz, Durand & Didon-Lescot, 1990). Several interacting mechanisms regulate the changes observed. They include change in water loss by evapotranspiration and associated modifications to hydrological pathways, differences in the trapping efficiency of water and atmospheric nutrient inputs, modifications to the soil profile, including the extent of litter accumulation, the degree of podzolisation and microbial activity, alterations in the extent of erosion and changes in snow accumulation and melt patterns. Changes in throughfall solute chemistry may also be important. These factors have been fully considered by Cresser & Edwards (1987). As might be expected, some of the greatest differences occur between forested and non-forested sites, as a consequence of the acidifying effects of tree growth on susceptible soils (Johnson *et al.*, 1991; Krug & Frink, 1983). In part these are attributable to the greater storage of base cations in the tree canopy, especially for species that are not deciduous.

Improved drainage and water quality

Soils may be drained for a variety of reasons. The most obvious one is because they are too wet for healthy plant growth, and soil aeration needs to be improved. From the discussion of oxidation–

reduction reactions in soils in Chapter 4, it should be clear that improved aeration resulting from drainage will lower the solubility of iron and manganese, reduce the extent of denitrification losses of gaseous nitrogen species, encourage nitrification, and encourage oxidation of reduced sulphur species to sulphate. These reactions liberate protons (see equations (4.23)–(4.26)), so drainage of waterlogged soils may generate a considerable degree of soil acidification.

The effect of these reactions upon drainage water chemistry may be acidification and increased mobilisation of nitrate and sulphate, with a decrease in soluble iron and manganese (in spite of the reduced pH). Soluble organic carbon may also decrease considerably with improved aeration.

As we saw in Chapter 4, drainage schemes may also be established to overcome salinity problems. These may arise in humid climates during reclamation of salt marshes, in which case the incoming water may be rainfall, or in arid climates as components of irrigation projects. The drainage water chemistry in the former case will slowly decline in salinity in general and sodium and chloride concentrations in particular. Salt water contains appreciable quantities of sulphates, and drainage of such marshes may be associated with marked falls in pH with increased aeration and oxidation. In arid climates, the drainage water chemistry depends upon the soil alkalinity and calcium carbonate content, and the irrigation water quality.

Agrochemicals, soils and freshwater quality

Most soils in agricultural use, especially if the agriculture is intensive, are treated with one or more agrochemicals from time to time. These include fertilisers and liming materials, and a range of pesticides which includes herbicides, fungicides, insecticides, molluscicides, etc. Heavy reliance is placed upon soils to retain these agrochemicals and their degradation products to avoid their transfer to potable water supplies at unacceptable concentrations. The purpose of this section is to examine briefly how successful soils are in fulfilling this role.

Since early in 1985, the pollution of water by agricultural practices has been covered by part II of the 1974 Control of Pollution Act in England and Wales (Joyce, 1986). The act includes a code of good agricultural practice which recognises the fact that potentially polluting materials must be spread in the environment, but aims to minimise potential damage, and legal liability of the farmer. Planned discharges are made after consultation with the relevant water authority.

Nitrogenous fertilisers and manures

There is a close link between nitrate concentrations in streamwater and intensity of agricultural activity (see also Chapter 7, and Walling & Webb (1981) and Whitehead (1990)). This does not simply reflect the rate of nitrogenous fertiliser use. Liming and crop removal produce conditions which favour mineralisation of organic nitrogen and nitrification. Moreover rainfall concentrations of nitrate tend to be higher in drier, lowland areas better suited to agriculture. Both in the atmosphere and in soils the dilution effect is reduced in such areas, as is any favourable contribution of denitrification to the overall nitrogen cycle.

There is often a pronounced seasonal pattern in agricultural catchments in the effect of a very heavy precipitation event upon riverwater nitrate concentration (Walling, 1980). In summer, the flush of water through the soil may be sufficiently rapid to reduce significantly the efficiency of nitrate uptake by plants, so that the concentration in the streamwater increases. In winter, on the other hand, plant nitrate uptake is much less significant, and a heavy storm tends to cause a simple dilution of nitrate concentration compared to that in baseflow water.

It has been pointed out that the organic nitrogen in an agricultural soil may be 150 times greater than the fertiliser nitrogen typically added annually by farmers (Dunning, 1986). Thus although in some areas heavy precipitation immediately following applications of highly soluble ammonium nitrate may lead to nitrate pollution episodes, this is only a minor component of the total problem. The bulk of the applied ammonium is retained on cation-exchange sites. Most farmers are far too cost conscious to apply fertiliser under such circumstances. Nevertheless, nitrate accumulation is occurring in some groundwaters, a fact which causes considerable concern because of the high cost of removal by current technology. It is always possible, of course, that developments in biotechnology may give rise to a cheaper removal technique in the long term.

Manures are often applied to land as a method of disposal and to increase soil organic matter and, particularly, nitrogen and phosphorus contents. Mineralisation and nitrification of organic nitrogen may contribute substantially to total pollutant load of adjacent drainage waters.

Other fertilisers

Unlike nitrates, phosphates are generally relatively insoluble. When they do dissolve, most of the phosphate leached down the soil profile is fixed by adsorption by soil mineral components (see Chapter 4). Thus the probability of serious pollution of freshwaters from normal use of phosphate fertilisers is slight. Pollution from sludge disposal or sewage outlets is more likely to be significant, and when it does occur may lead to the growth of algal blooms. Phosphate and nitrate are often the key nutrients limiting plant growth in oligotrophic lakes (Higgins & Burns, 1975). When the two pollutants occur together, growth may become very rapid, and cultural eutrophication is said to take place. Mobilisation of potassium from potassium fertilisers is rarely regarded as a source of pollution. Most of the element is retained on cation-exchange sites or fixed by clay minerals as it moves down the soil profile.

Liming materials

In most agricultural soils, liming is of no consequence to water quality except insofar as it provides optimal conditions for ammonification and nitrification of organic nitrogen. In uncultivated upland catchments, as we saw earlier in this chapter, it may have beneficial effects by reducing water acidity and increasing its calcium content.

Pesticides

The fraction of any applied pesticide dose or its degradation products that eventually reaches groundwater or a stream is the net result of a number of interactive processes (Boesten, 1987). These include the rates of chemical transformations in the soil, uptake by plants and soil organisms, the distribution between the soil gas, liquid and solid phases, and transport rates through the soil in the gas and liquid phases. Transport rates in soil solution depend upon adsorption coefficients on soil solid components, the excess of precipitation over evapotranspiration, and crop cover and growth stage. The movement of pesticides into rivers has not been studied as fully as it might have been over recent years, partly because of analytical difficulties at the very low concentrations involved, although this situation is now improving. A good insight into the techniques used to study the mobility of pesticides in soils may be obtained from Volume 2 of the *Proceedings of the British Crop Protection Conference* (*Weeds*) in 1987, published by the British Crop Protection Council.

7

Soils and pollution

From the dawning of pre-history, the human race has used soil as a depository for the waste products of its activities. Excrement, human and animal remains, ashes, crop and food residues and the superfluous scrap of its low-tech industries would be dumped in pits or simply left to lie on the surface. There the detritus would be subject to microbial and chemical decomposition, depending upon what it was. At this stage the boundary between pollution and natural biogeochemical recycling was not particularly distinct. Man was just one of many animal species producing waste materials. One major difference between humans and other animals, however, lies in the scale in which they tend to congregate into communities, thus generating zones with very substantial quantities of waste for disposal. A second divergence comes from the ability to use tools, which massively escalates the capacity for refuse production. One of these tools, fire, may lead to contamination of the atmosphere, as well as to solid residues for disposal. Thus it is the scale of human activities, rather than their fundamental nature, which leads to what are generally conceived as pollution problems.

Most people regard pollution as the result of subjecting ecosystems to loads of waste which are so great that they interfere perceptibly with the quality of the environment. Obnoxious smells in the air, tainted water supplies, visibly damaged trees and crops, animals or humans suffering from toxicological problems, or visible waste accumulation fall into this category. Such sensory manifestations are invariably the result of gross pollution rather than just pollution, however.

To the environmental scientist, some types of pollution may be regarded as serious, and requiring remedial action, long before their presence may be detected by human senses. Pollution is a term used to describe the presence of any elemental, ionic or molecular species at a concentration which has been accidentally raised as a consequence of human activity. Thus it may, but does not necessarily, pose a problem to human, animal or plant health, and its detection and quantification may require sensitive and selective methods of chemical analysis.

Sources of soil pollution
Sludges and animal wastes

Human and animal excretions remain a major source of soil pollution, although over much of the world sewage is treated before the somewhat less unpleasant residues are distributed over land (Lester, 1990). The risk of contamination of potable water supplies by pathogenic micro-organisms from either source is a cause of particular concern, either where surface runoff into streams or lakes used to provide the water is possible, or where wells are used in remoter areas and the soil is very freely draining to the water table.

In industrial regions, sewage sludge may contain substantial concentrations of heavy metals, as well as significant amounts of synthetic organic chemicals, depending upon the processes being used in the area. Much interest has focussed upon the presence of toxic heavy metals, especially cadmium, chromium, copper, mercury, nickel, lead and zinc, either because of their common occurrence or because of their toxicity at quite low concentrations (Sauerbeck, 1987). Heavy metal concentrations are commonly one to two orders of magnitude greater in sludge than in unpolluted soil.

Clearly there is a potential conflict of interest between the fertiliser value of sludge and soil pollution. In Europe, agricultural use of sludge ranges from as low as 4% in Ireland up to around 60% in Switzerland (Sauerbeck, 1987). To some extent this reflects both the allowed concentrations in agricultural soils and sludges, which vary from country to country, and the relative amounts of 'safe' sludge available. Allowable concentrations may be adjusted according to soil type; for example, copper, nickel and zinc may be potentially less toxic on alkaline, calcareous soils, and higher limits may be set for such soils.

It may be highly desirable in some areas to increase the organic matter content of soils, perhaps to increase water holding capacity or cation exchange capacity of sandy soils (see Chapters 3 and 4). Certainly the production of sludge is on a massive scale, so beneficial, or at least safe, disposal methods are in great demand. The Federal Republic of

Germany, for example, produced *ca* 45 million tonnes of sludge containing over 1 million tonnes of organic matter annually in the late 80s (Sauerbeck, 1987). In the UK in the 1970s, sludge contributed about 200 000 tonnes of phosphorus to soil each year, about the same amount as imported fertilisers.

Organic solvents, particularly chlorinated solvents (Crathorne & Dobbs, 1990), may also pose problems, especially if they pass through soil and reach potable water supplies. The half life of oil and grease in soil in one study in Oklahoma exceeded a year (Loehr, Martin & Neuhauser, 1992). For alkanes and naphthalenes the half life was generally less than 30 days. The application of oily sludge increased the soil temperature by 1–5 K, and raised the pH of acid soils by up to 1 pH unit.

Fertilisers

Just as sewage sludges spread for their fertiliser effect may contain undesirable pollutant element concentrations, so too may commercial agricultural fertilisers. For example, phosphates may contain undesirably large amounts of cadmium. Consideration must therefore be given to the possible consequences of regular, long-term additions of such materials to soils and to the human food chain. Figure 7.1, for example, shows the accumulation of exchangeable cadmium near the surface of two arable soils in Egypt (El-Sayad, 1988). Exchangeable cadmium concentrations were typically one order of

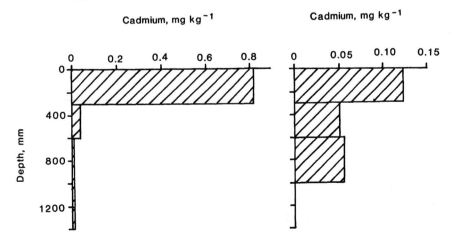

Fig. 7.1 Effect of cadmium accumulation at the surface of two arable soils in Egypt on exchangeable cadmium versus soil depth profiles.

magnitude lower at the surface of nearby uncultivated soils. Care is needed in the interpretation of such data, to make sure that the observed effect is not simply the consequence of a redistribution of native soil cadmium to increase the exchangeable cadmium near the surface.

Most members of the general public tend automatically to think of pollution of drinking water when fertilisers are mentioned in the context of pollution. This aspect has already been considered in Chapter 6. Often the link between use of nitrogenous fertiliser and water quality in a region is all too obvious. Figure 7.2, for example, shows the relationship between the amount of agricultural land in the catchments upstream from the sampling points and the associated riverwater nitrate concentrations in two British rivers (Edwards *et al.*, 1990). The increase in nitrate flux leached into the rivers from arable land is immediately apparent. It should be noted that the nitrate leaching must be associated with additional base cation leaching, and therefore the nitrogenous fertiliser is effectively a pollutant which damages the soil in the long term.

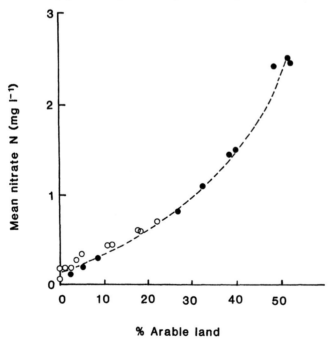

Fig. 7.2 Effect of amount of arable land (and hence fertiliser nitrogen use) upon riverwater nitrate concentration (mean for 1980–6) for the River Dee (open circles) and Don (filled circles) in Scotland.

Sources of soil pollution

Atmospheric inputs

Substantial inputs of pollutants reach the soil from the atmosphere. Those which attract most attention are radionuclides, from testing of nuclear weapons or accidents at nuclear power stations, oxides of sulphur and nitrogen and the associated sulphuric and nitric acids (acid rain), ammonia, lead from car exhausts, and lead and other heavy metals from industry. Heavy metals may move through the atmosphere over very large distances. Figure 7.3, for example, shows how lead concentration has increased in the surface horizons of a series of relatively remote mature forest sites in north-east Scotland over a sampling interval of 39 years (Billett, Fitzpatrick & Cresser, 1991). Conclusive evidence for increasing atmospheric pollution with these elements comes from the analysis of fractionated polar ice cores, and from examination of recent changes in trace metal concentrations with depth of dated sediment cores (see, for example, Reuberg *et al.* (1990) and Rippey (1990)).

Radionuclide deposition

The accidental release at Chernobyl in the spring of 1986 of long-lived radionuclides into the environment highlighted the need to understand the chemical behaviour of radionuclides in the soil–plant system.

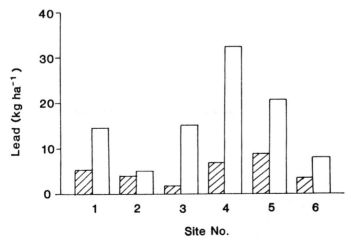

Fig. 7.3 Lead accumulation in surface organic soils at several sites in the Alltcailleach Forest in the Grampian Region of Scotland. Unshaded bars show lead contents in 1987, shaded bars the contents in 1949/50.

The behaviour of radionuclides in soil depends both on the isotopes involved and upon their concentration, as well as on the physicochemical nature of the recipient soil. Of particular concern after Chernobyl were the radionuclides ^{90}Sr and ^{137}Cs. The former is a beta emitter with a half life of 28 years, the latter a beta and gamma emitter with a half life of 30 years. Strontium is a divalent metal cation and behaves similarly to calcium in soil. Caesium is monovalent, and behaves like potassium in soil, although its atomic radius is slightly greater than that of potassium.

Movement of the radionuclides ^{90}Sr and ^{137}Cs through the soil will largely depend on the cation-exchange capacity, the relative amounts of diverse cations on the exchange sites, and on the charge and hydrated ion radius of the radionuclide (see Chapters 2 and 4). Divalent cations such as strontium will tend to be more tightly held on exchange sites than monovalent cations like caesium.

In Britain, substantial amounts of the radionuclides were deposited on highly organic, acid upland soils, many of which were used for rough grazing. The possibility of the radioactive caesium or strontium entering the food chain was therefore a cause for concern. The high cation-exchange capacities and generally low base saturation of the acid upland soils favoured exchange between the radionuclides and protons, especially for ^{90}Sr. Caesium tended to be more mobile, particularly where there was competition from ammonium from animal wastes.

In organic soils, radionuclides may be held both by simple charge attraction and as a result of the formation of insoluble organic chelates. In more mineral soils, however, where cation exchange capacity is largely clay-based, radionuclides tend to be held by charge only. Clays often hold radionuclides, particularly caesium, with great strength. This is because caesium becomes trapped within the clay lattice, preventing subsequent ion exchange with the soil solution (Tamura & Jacobs, 1960). This situation is analogous to potassium fixation in 2:1 type layered aluminosilicates (e.g. micas, as discussed in Chapter 2). In fact, ^{137}Cs is so tightly held in mineral soils that its distribution may be used as a tracer for quantifying soil erosion on arable fields (Quine & Walling, 1991; Walling & Quine, 1991).

Because strontium and caesium are analogues of calcium and potassium respectively, they are highly 'bioactive', and are readily taken up by plants, as well as soil microbes and animals. Plants and microbes whose growth is limited by calcium and/or potassium will tend to take up larger amounts of strontium and/or caesium. Clearly therefore the behaviour of a pollutant radionuclide in soil results from the complex interaction of both the chemical and the biological properties of the radionuclide and the soil environment.

Ammonia

Over recent years the deposition of ammonia to soils has attracted considerable attention, especially in areas such as the Netherlands where animal production is very intensive (van Breemen, Boderie & Booltink, 1989). Although ammonia is a weak base, and much ammonia is codeposited with sulphuric acid as ammonium sulphate, it contributes substantially to soil acidification if it is nitrified:

$$NH_4^+ + 2O_2 \rightarrow NO_3^- + H_2O + 2H^+$$

Following a study in Sweden, Witter (1991) has even suggested that it could be worthwhile to add calcium chloride along with poultry farm waste to minimise ammonia volatilisation. The calcium precipitates HCO_3^- as calcium carbonate, thus lowering the pH:

$$Ca^{2+} + HCO_3^- \rightarrow CaCO_3 + H^+$$

Because of the acidifying effect which they have as a consequence of nitrification, it is increasingly common to consider ammonia and ammonium deposition as constituents of acid rain.

Soils and acid deposition

Inputs of nitric and sulphuric acids to soils result in a shift in ion exchange equilibria in the short term, and thus to enhanced leaching of base cations (see Chapter 4). If the base cations leached from cation-exchange sites during a rainstorm are replenished subsequently by geochemical weathering (see Fig. 4.4), the pH of the soil will not fall. Thus soil may be strongly buffered against pH change for many years.

It is all too easy to reach the conclusion over this period that acid deposition is having no adverse effect on soil. Johnson and Lindberg (1989), discussing the results of investigations of possible acidification of forest soils in the Walker Branch watershed of East Tennessee, where little acidification was apparent, pointed out that substantial change might eventually be observed if observations continued for long enough. If, on the other hand, a soil is naturally low (or already highly depleted) in readily weatherable minerals, the base saturation and pH of the soil will fall rather more rapidly in response to deposition acidification, until a new equilibrium is reached. The extent to which this happens at various depths down the soil profile may be influenced by cation uptake and geochemical recycling, as is the rate of acidification. Drainage water will be more acidic too, and will contain more aluminium and less calcium, the amount of aluminium and acid mobilised depending upon the total mobile anion content of the drainage water. The mobilisation of cationic monomeric aluminium by acidic deposition has been clearly demonstrated by Driscoll, Johnson, Likens & Feller (1988), who compared the chemical properties of water draining the very extensively studied Hubbard Brook catchment in New Hampshire with that in a less polluted catchment in British Columbia with soils derived from similar glaciated granitic bedrock.

For forests at higher altitudes in any given region, acidic deposition may be appreciably greater than at lower altitudes. However Johnson, Siccama, Silver & Battles (1989) suggested that there was no conclusive evidence that soils in the high elevation forests that they studied in New York and New Hampshire had been acidified by acid deposition. In the authors' experience it is always necessary to consider acid deposition effects in relation to soil parent material. For a regional survey in the UK, while moorland podzol B horizon soils had acidified considerably where the soil had derived from quartzite and old, acidic sandstones or granites, no such acidification was observed for more base-rich rock types (Bull *et al.*, 1992).

The adsorption of sulphate by hydrous oxides of iron and aluminium in acid soils was mentioned in Chapter 4 when the sulphur cycle was discussed. This anion adsorption reaction may be represented by the equation:

$$\text{soil}\!=\!(OH)_2 + SO_4^{2-} \rightleftharpoons \text{soil}\!=\!SO_4 + 2OH^- \qquad (7.1)$$

In a pH-buffered soil, the liberated hydroxide ions are immediately neutralised by hydrogen ions. To maintain the activity of H^+, more H^+ ions must pass into solution from cation exchange sites, and an equivalent amount of base cations must be adsorbed in their place. Thus sulphate adsorption is associated with base cation retention. Sulphate saturation, on the other hand, is associated with base cation leaching and reduction in soil pH.

Figure 7.4 shows the change in pH at different depths down a soil profile under mature coniferous forest in northeast Scotland (Billett, Fitzpatrick & Cresser, 1988). In this instance the soils have acidified partly at least as a consequence of their ability to adsorb sulphate being fully used up. Examination of the sulphate adsorption characteristics of the lower, hydrous oxide-rich mineral horizons of this profile showed some sulphate adsorption in 1949, but substantial sulphate leaching in 1987 (Billett, Parker-Jervis, Fitzpatrick & Cresser, 1990).

Leaching experiments with simulated acid deposition and reconstituted soil cores containing hydrous oxide-rich horizons show increased leaching of base cations as soon as sulphate saturation occurs. Figure 7.5 shows a set of results for one such experiment conducted in the authors' laboratory with simulated pH 3.5 rainfall over several weeks. Calcium concentration rises with the sulphate concentration in the drainage water, as the soil becomes progressively more sulphate saturated. The sulphate concentration in the 'rainfall' was constant throughout.

The fact that sulphate adsorption and an eventual increase in the degree of sulphate saturation are frequently precursors to soil acidification means that there is sometimes a link between the pH of iron- and aluminium-rich B horizons of podzols and their capacity to adsorb sulphate. Michopoulos (personal communication) has demonstrated such a relationship for Scots pine (*Pinus Sylvestris*) stands in Northeast Scotland (Fig. 7.6).

For deep peat soils, there is no possibility of mineral weathering replenishing leached base cations, and therefore the pH of surface peats is bound to fall if the acid concentration of precipitation increases while other factors remain unchanged. Conversely the peat pH rises if the concentration of acid in the precipitation falls, as a consequence of base

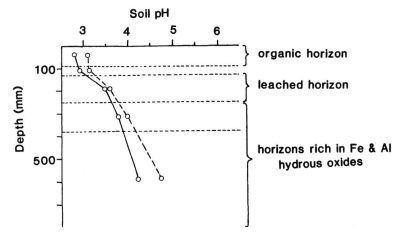

Fig. 7.4 Typical example of the change in pH versus depth of a podzol profile under mature Scots Pine (*Pinus Sylvestris*) forest in northeast Scotland between 1949/50 (dashed line) and 1987 (full line).

160   Soils and pollution

cation retention from the precipitation (Skiba & Cresser, 1989). This may be very simply demonstrated by simulation experiments, and some typical results from such an experiment are shown in Fig. 7.7. A shake-and-centrifuge technique has been used to follow the equilibration of the peat with 'rain' at pH values of 3.5, 4.5 and 5.5. Both acidification and recovery are illustrated. Experiments of this type may be used to predict the effect of precipitation acidity upon peat and associated drainage water pH values (Skiba & Cresser, 1989). The relationship between peat pH and acid deposition has been verified by field observation (Skiba, Cresser, Derwent & Futty, 1989).

In the case of nitric acid in precipitation, as mentioned in Chapter 6, direct uptake of nitrate by plants is possible, especially on soils of low nitrogen status. This uptake consumes H^+ ions:

$$ROH + NO_3^- + H^+ \rightarrow RNH_2 + 2O_2 \tag{7.2}$$

This reaction temporarily offsets the effects of H^+ input. However, the reverse reaction, mineralisation of organic nitrogen followed by nitrifi-

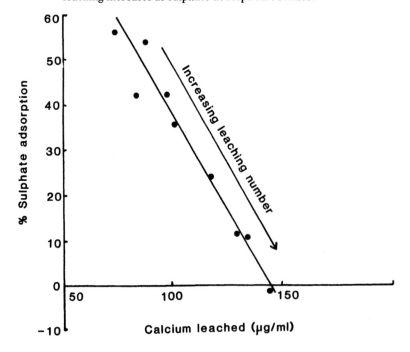

Fig. 7.5 Relationship between sulphate adsorption by a reconstituted soil profile and calcium concentration in leachate from the profile when the soil was subjected to simulated rainfall at pH 3.5 over several weeks. Calcium leaching increases as sulphate adsorption declines.

cation, liberates the H^+ ions again. Leaching of soluble organic nitrogen compounds, which may be very considerable in acid forest soils (Stevens & Wannop, 1987), is thus a mechanism for amelioration of nitric acid deposition effects.

The acidification of soils in potentially sensitive ecosystems is of widespread interest, and numerous attempts have been made to model acid deposition-driven acidification. Detailed discussion of the approaches used is outside the scope of this short volume, but a concise chapter on the application of two of the better known models, RAINS and SMART, to historical soil chemistry data from Sweden provides a useful introduction to this approach (Posch, Falkengren-Grerup & Kauppi, 1989). As should be clear from Chapters 3 and 4, the nitrogen

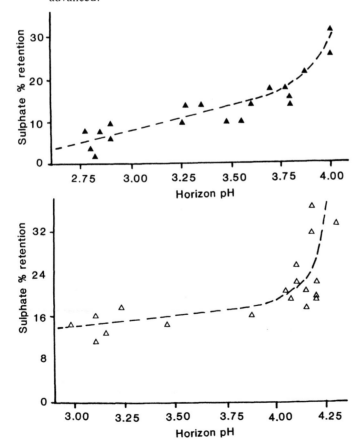

Fig. 7.6 Relationships between soil ability to adsorb sulphate and pH at two depths for 19 forest sites in northeast Scotland. The open triangles are for soils at greater depth, where sulphate saturation and acidification are less advanced.

162   Soils and pollution

cycle is complex, which makes reliable quantitative prediction of the acidification of soils by nitrogen deposition extremely difficult (Malanchuk & Nilsson, 1989). Moreover, the situation is further complicated by the possible interactive effects of nitrogen compounds and other atmospheric pollutants (Wellburn, 1988).

Towards the end of the 1980s, there was much interest throughout Europe and in North America in mapping the maximum amounts of pollutants which could be tolerated by soils and freshwaters without damage being done, the so-called 'critical load', an approach first widely recommended in Scandinavia (Nilsson, 1986). At the time of writing this approach is being extensively employed, and the resulting maps are being used to plan pollution abatement strategies. However, there are still doubts as to the precision with which critical loads may be quantified, partly because of uncertainties about mineral weathering rates under field conditions, and partly because of disagreement about the allowable accumulation of organic matter (and thus of organic nitrogen – a proton sink).

Effects of increasing carbon dioxide

The concentration of carbon dioxide in the atmosphere has been rising exponentially since the early eighteenth century (Woodward, 1988). The carbon cycle, like the nitrogen cycle, is highly complex, and

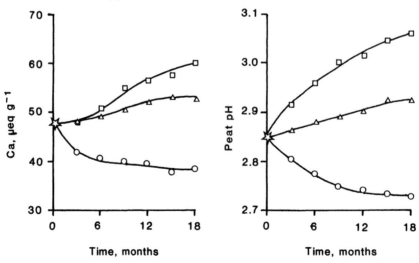

Fig. 7.7 Effect of simulated rain at pH 3.5 (circles), pH 4.5 (triangles) and pH 5.5 (squares) on the calcium content and pH of an upland hill peat from a moderately polluted site in the UK.

is also interactive with the nitrogen and sulphur cycles. Quantitative information on probable effects of increasing atmospheric carbon dioxide concentrations upon soil chemistry and soil microbial activity are woefully lacking at the time of writing. The nature of the complex set of interactions which will need to be considered is discussed briefly in Chapter 8.

Adsorption of pesticides by soils

The past few decades have seen increasing use of a wide range of pesticides in agriculture, which ultimately find their way directly or indirectly into soils. The risk of these compounds or their metabolites finding their way into the human food chain *via* drinking water has given rise to a need for extensive research into pesticide mobility in soils.

The phase distribution of pesticides in a soil is determined by the nature of adsorptive soil colloids, the physicochemical characteristics of the pesticide, and a host of physical, chemical and biological factors which are determined by the particular soil environment. Of the properties of the pesticide, the most important are the electronic structure and the water solubility. The readiness with which a molecule ionizes is the main reason why pesticide adsorption and movement in soil is strongly pH dependent.

The soil components which control adsorption are the clay particles, amorphous mineral matter and soil organic matter (see, for example, Dios Cancela, Romero Taboada & Sanchez-Rasero (1992)). These components are not always spatially separate; soil aggregates are often an intimate association of all three adsorptive components (see Chapter 3). Pesticide adsorption in soils is not a simple, single process, therefore, although in some soils one particular process may predominate. As an indication of the complexity of the problem, it is worth elaborating briefly on the nature of the adsorption processes that are important. In the interests of brevity, the authors have concentrated the discussion upon adsorption by soil organic matter.

Types of pesticide adsorption to soil organic matter
Van der Waals' forces
These are important for neutral polar or non-polar molecules and are weak physical attractions which result from short range dipole–dipole or induced dipole–dipole interactions. These forces are additive, and may result in considerable attraction for large molecules.

Hydrogen bonding

Hydrogen bonding is a type of dipole–dipole interaction in which hydrogen atoms serve as a bridge between two electronegative atoms, one being held by a covalent bond and the other by electrostatic forces. The presence of oxygen-containing functional groups and amino acid groups in organic matter indicates the possibility of hydrogen bonding with organic pesticides containing similar groups. Hydrogen bonding is limited to acid conditions where organic matter carboxylic acid (—COOH) groups are not dissociated. The forces of attraction are weaker than ionic, but stronger than van der Waals' forces.

Hydrophobic bonding

Adsorption by hydrophobic bonding involves non-polar organic compounds, or compounds with relatively large non-polar parts. Adsorption occurs onto the hydrophobic surfaces of organic matter, which include fats, waxes and resins. The solute usually has low solubility and water will not compete at the adsorption site.

Ion exchange

Adsorption by ion exchange applies to cationic pesticides (e.g. Paraquat) and to weak basic compounds (e.g. s-triazines), which can become cationic by protonation. Adsorption occurs by cation exchange processes to the —COOH and —OH groups of the organic matter. Permanently charged cationic pesticides involve simple cation exchange processes. It is possible, however, that some organic matter sites may not be physically available to large organic cations, but available to smaller inorganic cations.

Charge transfer

Charge transfer mechanisms are electrostatic interactions which occur when electrons are transferred from an electron-rich donor, such as some pesticide, to an electron-deficient acceptor such as organic matter. Free radical intermediates and quinone-like structures are possible electron acceptor sites in organic matter. Charge transfer interactions occur only within very short distances.

Ligand exchange

Adsorption by this mechanism involves replacement of one or more ligands by the pesticide. The adsorbate molecule must be a stronger chelating agent than the replaced ligands. The adsorbate molecule may displace water of hydration acting as a ligand. Figure 7.8 illustrates ligand bonding of a pesticide to soil organic matter.

Chemisorption

This involves formation of a covalent chemical bond between the adsorbate and the adsorbent. Chemisorption is an exothermic process and is characterised by considerable heat of adsorption. Adsorption by this mechanism can take place at very low adsorbate concentrations and still produce adsorbent site saturation.

The main mechanism by which organic cationic pesticides adsorb to soil particles is ion exchange. Acidic particles, which ionise in soil solution to produce anionic species, tend to bind to soil organic matter *via* polyvalent cationic bridges or, at low pH, *via* protonated groups such as amino groups. Basic pesticides which ionise to form cations in soil solution will readily adsorb directly to soil organic matter. The basic pesticides will tend, therefore, to be more strongly held in most soils than the acidic pesticides, and will be less likely to leach into groundwater.

The pH dependence of pesticide adsorption is well documented. The maximum adsorption of each pesticide will tend to take place at soil pH values close to the pK value for adsorption (Weber, Weed & Ward, 1969). Increased adsorption of a pesticide reduces the active concentration in soil solution, and therefore generally reduces the pesticidal effect.

Most of the pesticides currently used in agriculture are non-ionic and

Fig. 7.8 An example of ligand exchange bonding of a pesticide to soil organic matter. The metal ion (M) functions as a bridge.

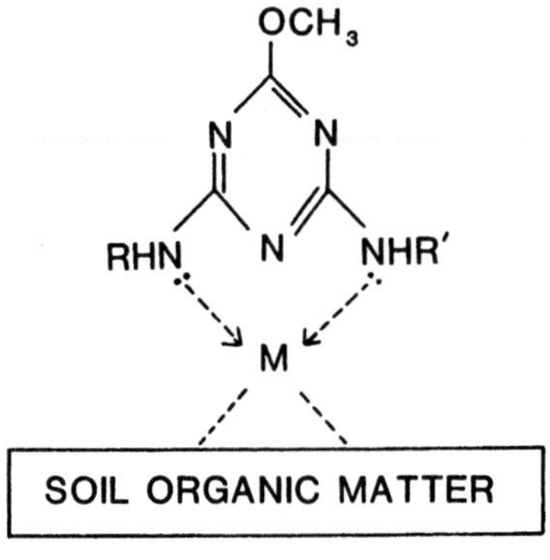

they will tend to be held by organic matter as a result of cation dipole and coordination bonds, hydrogen bonds, and van der Waals' forces.

Organic matter fractions involved in pesticide adsorption
While the quantitative effect of soil organic matter on pesticide adsorption is well documented, far less is known of the qualitative aspects. It is the humic fraction of the soil organic matter (see Chapter 3) that appears to have the highest affinity for most pesticides. Dunigan & McIntosh (1971) fractionated soil organic matter and found little adsorptive capacity in material from hot water extracts (polysaccharides) and ether extracts (waxes, fats, resins, etc.), but high adsorptive capacity in the humic acid and lignin-associated fractions. It appears that carboxyl, phenolic hydroxyl and carbonyl groups existing as quinone groups are important for the adsorption of pesticides by humic acids (Li & Felbeck, 1972).

Perhaps the most conclusive evidence to date of the importance of humic components of soil organic matter in pesticide adsorption comes from field studies using ^{14}C-labelled pesticides, in which ^{14}C label searches allow a budget of pesticide distribution with time to be characterised. More than 80% of ^{14}C-labelled atrazine applied to soil remained in the soil after nine years (Capriel, Haisch & Khan, 1985). Of this ^{14}C, only 10% was associated with non-humic components, with the remaining 90% associated with the humic components, humic acid, fulvic acid and humin.

An understanding of the dynamics of pesticides in soil, particularly in relation to adsorption to soil organic matter, eventually will enable us to predict the likely movement of pesticides into groundwater, a topic of considerable current concern. Qualitative and quantitative understanding of the factors which regulate the fate and mobility of pesticides in soils, and their transfer to drainage water, is an important prerequisite to being able to provide advice on the safe use of these potentially lethal compounds. It may not be too long before pesticide application rates are regularly tailored to the characteristics of the soil types present in the area being treated (Vink & Robert, 1992).

This concept of land management to minimise pollution of either the atmosphere or, more commonly, of groundwaters or surface waters, is one which is likely to attract considerable attention over the coming decade. For example, Mitchell and McDonald (1992) have discussed the moorland management implications of discolouration of water by peat following periods of rain after drought. For many soils, however, contamination is already a serious problem. The methods used to deal with contaminated soils warrant a book of their own. However, the beginner can gain some insight into this topic by reading the excellent concise review by Bridges (1991).

8

The future of soil chemistry

Conscientious readers of the preceding seven chapters should, at this stage, have a reasonable grasp of the fundamental principles of soil chemistry and should be well aware of its importance in the context of environmental science as a whole. They might well be left asking themselves: 'Where does the subject go from here? What, if anything, do we still need to do or to find out?' Those whose studies or interests are motivated by career prospects might be wondering what the future holds in store for aspiring soil chemists. With this possibility in mind, the authors thought it would be appropriate to bring the book to a close by taking a brief look into the future, while at the same time seeing what lessons might be learned from past progress and mistakes.

The changing nature of research in soil chemistry

One recurring concept throughout this short volume has been that of the dynamic nature of soils. Soil scientists have been aware of this fundamental idea for many decades, and it is central to discussions of soil evolution in both the presence and absence of anthropogenic influences. Much of our understanding has been based on qualitative, or at best semi-quantitative, understanding of the fundamental chemical, physical and biological mechanisms involved in soil development and change. Over the past few years there has been an increasing tendency to try to put this understanding on a more quantitative basis. As a result we are now seeing soil chemists involved in attempts to make reliable mid- and long-term predictions of rates of soil change in response to the imposition of various external pressures. These pressures include, for example, effects of prolonged heavy cropping, consequences of changes in land use and management practices, response to climate change or atmospheric pollutant deposition loads, and the introduction of microbial inocula, including genetically engineered microorganisms (GEMs). Soil chemists are also involved in the assessment of the suitability of soils for disposal of sludges and other wastes, an area where understanding of sorption by soil constituents is particularly important (Schulthess & Sparks, 1991).

Modellers sometimes tend to start, for simplicity, on a very local scale, and then to expand their projections progressively to larger and larger areas. Such attempts, however, often expose a lack of sufficient quantitative knowledge of fundamental chemical and other properties of soils. One of the most central examples of such a problem is our limited knowledge of the kinetics of geochemical weathering rates and reactions *under field conditions*, processes right at the hub of soil change. We need to know what soil atmosphere and soil solution changes occur in response to changed pollutant loads or changed biota at the level of the individual soil particle or even particle surface. Adsorption of the microbial component of the soil biota involves surface chemistry at the molecular level. We then need to know what effects these changes have upon the release of major and trace nutrient elements from soil minerals and upon the fate of the elements released. It is then necessary to be able to relate what happens at the micro scale to what happens on a much larger scale. In the authors' opinion we are still a long way from being able to do this with any degree of confidence.

An alternative approach to modelling adopts the stance that the soil–plant–water system is too complex to consider quantitatively at the fundamental process level. Whole series of processes are considered to be represented by single equations (i.e. the processes are 'lumped' together), the soil, in effect, being treated simply as one or more black boxes. Such an approach is often very useful for predictive purposes once the model has been set up and adequately calibrated. Great care is necessary, however, if predictions are made outside the range over which the model has been effectively calibrated.

Soil as a component of the whole ecosystem

Increasingly over the past decade there has been a tendency to treat soil as part of a whole ecosystem, rather than as an isolated entity. This change in emphasis has partly arisen because of the central role which soil plays in both plant growth and the regulation of drainage water quality. A growing awareness on the part of environmental scientists of this central role of soil has resulted in their needing more information on soil chemical change with time. This has thrown a challenge at soil chemists to make more reliable long-term predictions of soil change. The soil chemist, in turn, has found the need to consider the soil as an ecosystem component to ensure that no crucial process is ignored.

A whole-ecosystem approach to soil chemistry is essential when considering some of the major problems facing the world today. The most striking example of this need is when considering the impact of increasing atmospheric carbon dioxide concentrations. Higher concentrations of carbon dioxide should stimulate photosynthesis and improve crop yields. This, in turn, will change plant nutrient demands from the soil. If the soil cannot meet the increased demands, deficiency problems may occur. Increased concentrations of carbon dioxide will also change the water use efficiency of plants; this may alter the kinetics of soil weathering. There will also undoubtedly be changes in plant root growth, which will, in turn, influence soil chemical and microbiological dynamics. As a result there may well be a change in the distribution and amount of carbon stored in soil, which would influence the carbon dioxide returned to the atmosphere, thus providing an active feedback mechanism. The situation is clearly very complex, and so far we have not considered the impact of climate change!

There is a growing awareness at the present time of the need to consider soil management not just in terms of production of high crop yields in the short term, but also in terms of long-term conservation of ecosystem resources. At present, emphasis tends to be put upon sustainable agriculture. In the longer term, we need to shift this emphasis towards sustainable ecosystems, and it is in this context that future land management practices should be carefully selected. The authors are of the opinion that this will provide a new focus for soil-related aspects of agronomic research. It must not be forgotten, however, that interest in soil fertility in many countries still concentrates, of necessity, upon optimisation of agricultural yields, while in others agricultural overproduction is perceived as a major problem.

New research tools in soil chemistry

However intellectually capable he or she may be, the soil chemist is always constrained to some extent by the finite power of the research tools available to tackle any given problem. Over the past three decades, two major areas of development have been crucial to the success of soil chemistry research; they are computing and instrumental chemical analysis. The former allows rapid processing and reprocessing of massive bodies of data, essential for mapping and solution of complex mathematical problems on a speedy timescale. The latter, and especially developments in analytical atomic spectrometry, have allowed chemical analysis at a cost and on a scale which would have been difficult to conceive of in the 1950s.

The high sensitivity and selectivity of atomic absorption spectrometry, first using flames, and later using furnace and hydride generation techniques, has had a particularly dramatic impact. More recently, the capability of inductively coupled plasma atomic emission spectrometry for simultaneous trace and major nutrient element determinations on small volumes of samples has further widened research horizons. This, in turn, has been surpassed by the development of inductively coupled plasma mass spectrometry (ICP-MS), which matches or sometimes surpasses furnace atomic absorption spectrometry in terms of sensitivity, but retains the power of multielement analysis (Potts, 1987, Jarvis, Gray & Houk, 1991).

Another exciting opportunity offered by ICP-MS is its capacity for stable isotope analysis. Hitherto only ^{15}N has been widely used in this respect, to help elucidate the complex interactive components of the nitrogen cycle. This is another example of where analytical development has been an essential prerequisite to progress in soil chemistry. Thermal ionisation mass spectrometry (TIMS) has also been used in recent years for isotope analysis. For example, by determination of different lead isotopes, it is possible to apportion soil lead between local geochemical weathering and atmospheric deposition.

One important consequence of the development of methods of trace inorganic analysis, which allow sensitive, rapid and highly selective analysis, is the ease with which speciation studies may now be carried out. For example, it is possible to determine water-soluble, exchangeable, organic-matter-bound, carbonate-bound, amorphous-oxide-bound and crystalline-oxide-bound fractions of a trace metal cation in soil (see, for example, El-Sayad, Cresser, El-Gawad & Khater, (1988)). Such speciation should allow a better understanding of the fate of pollutants, or of the influence of cultivation and fertilisation techniques on trace element availability to plants. It must be stressed, however, that the results of such speciation studies are operationally defined (Tessier & Campbell, 1991). It is imperative always to bear in mind possible redistribution of determinant by side reactions which may only occur as a consequence of the fractionation steps being employed. A sound understanding of soil chemistry is therefore crucial.

In the elucidation of soil organic matter structure, progress has been less remarkable, but this largely reflects the complexity of the problem. Mention was made in Chapter 3 of the use of magic angle nuclear magnetic resonance spectrometry (NMR).

Mention should also be made here of the potential of remote sensing for assessment and mapping of soil properties and land use changes. There are exciting opportunities for large scale monitoring of crop and soil changes and crop health.

At a more fundamental level, it is interesting to note that bioassays are making a come-back. The analysis of plant material has been used for many years to diagnose nutrient disorders. Recently, however, there has been increased interest in the use of growing plants as indicators of soil nutrient availability. Two approaches have been adopted which address either the availability of nitrogen, phosphorus and potassium (Harrison & Dighton, 1990) or the phytotoxicity of ions (e.g. Al) in acid soil solutions (Aitken, Moody & Compton, 1990). Both methods are based upon the short-term exposure of the plant root system, and show excellent diagnostic potential.

Soil chemistry and policy makers

From the earliest days of pre-history, soil fertility has been a driving force behind policy decisions of community leaders. In those days the decision to move on and cultivate new areas might have to be made as a consequence of food production no longer being sustainable.

In more recent times soil chemists have played major roles in the successful development of agricultural and land use policy and national prosperity in many countries. In the process they have contributed unintentionally to many of the pollution problems which the modern world has to face. Currently they have an opportunity to redeem themselves for past inadequacies by finding the long-term solutions to these same pollution problems. To do so, they must always remember to consider soil as part of a delicately balanced global ecosystem.

REFERENCES

Chapter 1

Columella, L.J.M. (*ca.* 60 AD) *Res Rustica*, translated by Ash, H.B., 1941. The Loeb Classical Library, Heinemann, London.
De Saussure, Th. (1804) *Recherches Chimiques sur la Vegetation*. Paris.
Liebig, von, J. (1840) *Chemistry and its Application to Agriculture and Physiology*, second edition, 1842, Playfair, L., ed. London.
Russell, E.W. (1987) *Soil Conditions and Plant Growth, Eleventh Edition*, Wild, A., ed. Longman, London.
Schloesing, Th. and Muntz, A. (1877) Sur la nitrification par les ferments organises, *Comptes Rendus des Seances de l'Academie des Sciences*, **84**, 301-3.
Senebier, J. (1782) *Memoires Physico-Chimiques sur l'Influence de la Lumiere Solaire pour Modifier les Etres des Trois Regnes de la Nature, et sur-tout Ceux de Regne Vegetal*. Geneve.
Varro, M.T. (ca. 37 BC) *Res Rusticae*, translated by Hooper, W.D., revised by Ash, H.B., 1934. The Loeb Classical Library, Heinemann, London.

Chapter 2

Bowen, H.J.M. (1979) *Environmental Chemistry of the Elements*, Academic Press, London.
Brady, N.C. (1990) *The Nature and Properties of Soils*, tenth edition, Macmillan, New York.
Brown, G., Newman, A.C.D., Rayner, J.H. and Weir, A.H. (1978) The structures and chemistry of soil clay minerals: Chapter 2 in *The Chemistry of Soil Constituents*, Greenland, D.J. and Hayes, M.H.B., eds, John Wiley & Sons, Chichester.
Edwards, A.C., Creasey, J., Skiba, U., Peirson-Smith, T. and Cresser, M.S. (1985) Long-term rates of acidification of UK upland acidic soils. *Soil Use & Management*, **1**, 61-5.
Fitzpatrick, E.A. (1980) *Soils: Their Formation, Classification and Distribution*, Longman, London.
Greenland, D.J. and Hayes, M.H.B. (1978) *The Chemistry of Soil Constituents*, John Wiley and Sons, Chichester.

Jackson, M.L. and Sherman, D.G. (1953) Chemical weathering of minerals in soils. *Advances in Agronomy*, **5**, 219–318.

Landon, J.R. (1991) *Booker Tropical Soil Manual*, paperback edition, Longman, London.

Mason, B. (1982) *Principles of Geochemistry*, fourth edition, John Wiley, New York.

Sposito, G. (1989) *The Chemistry of Soils*, Oxford University Press, Oxford.

Walling, D.E. and Webb, B.W. (1981) Water quality: Chapter 5 in *British Rivers*, Lewin, J., ed., George Allen and Unwin, London.

Chapter 3

Achard, F.K. (1786) Chemische Untersuchung des Torfs. *Chemische Annalen fur die Freunde der Naturlehre von L. Crell*, **2**, 391–403.

Anderson, G. (1982) Recent observations on humus composition and properties. *Promocet*, 3–10.

Billett, M.F., FitzPatrick, E.A. and Cresser, M.S. (1988) Long-term changes in the acidity of forest soils of North-East Scotland. *Soil Use and Management*, **4**, 102–7.

Bohn, H.L. (1976) Estimate of organic carbon in world soils. *Soil Science Society of America Journal*, **40**, 468–70.

Bolin, B., Degens, E.T., Kempe, S. and Ketner, P. (eds.) (1979) *The Global Carbon Cycle*, Wiley, New York.

Cheshire, M.V. (1979) *Nature and Origin of Carbohydrates in Soils*. Academic Press, London.

Cheshire, M.V., Sparling, G.P. and Mundie, C.M. (1983) Effect of periodate treatment of soil on carbohydrate constituents and soil aggregation. *Journal of Soil Science*, **34**, 105–12.

Cooke, R.C. & Rayner, A.D.M. (1984) *Ecology of Saprophytic Fungi*, Longman, London.

Crasswell, E.T. and Waring, S.A. (1972) Effect of grinding on the decomposition of soil organic matter – 1. The mineralisation of organic nitrogen in relation to soil type. *Soil Biology and Biochemistry*, **4**, 427–33.

Flaig, W. and Sochtig, H. (1964) Einfluss organischer Staffe auf die Aufnahme anorganischer Ionen. *Agrochimica*, **6**, 251–64.

Fogel, R. 1983. Root turnover and productivity of coniferous forests. *Plant and Soil*, **71**, 75–85.

Forster, S.M. (1981) Aggregation of sand from a maritime embryo sand dune by microorganisms and higher plants. *Soil Biology and Biochemistry*, **13**, 199–203.

Greenland, D.J., Rimmer, D. and Payne. D. (1975) Determination of the structural stability class of English and Welsh soils, using a water coherence test. *Journal of Soil Science*, **26**, 294–303.

Greenwood, D.J. (1975) Measurement of soil aeration: in *Soil Physical Conditions and Crop Production*, MAFF Bulletin, 29, 261–72.

Hatcher, P.G., Maciel, G.E. and Dennis L.W. (1981) Aliphatic structures of humic acids. A clue to their origin. *Organic Geochemistry*, **3**, 43–8.

Jenkinson, D.S. (1965) Studies on the decomposition of plant material in soil. 1. Losses of carbon from ^{14}C labelled ryegrass incubated with soil in the field. *Journal of Soil Science*, **16**, 104–15.

Jenkinson, D.S. and Ladd, J.N. (1983) Microbial biomass in soil – measurement and turnover: in *Soil Biochemistry*, Vol. 5, Paul, E.A. and Ladd, J.N., eds, Marcel Dekker, New York.

Jenkinson, D.S. and Rayner, J.H. (1977) The turnover of soil organic matter in some of the Rothamsted classical experiments. *Soil Science*, **123**, 298–305.

Kirk, T.K., Connors, W.J., Bleam, R.D., Hackett, W.F. and Zeikus, J.G. (1975) Preparation and microbial decomposition of synthetic ^{14}C lignins. *Proceedings of the National Academy of Science (USA)*, **72**, 2515–19.

Kirk, T.K. and Fenn, P. (1982) Formation and action of the ligninolytic system in basidiomycetes. In *Decomposer Basidiomycetes: Their Biology and Ecology*, Frankland, J.C., Hedger, J.N. and Swift, M.J., eds, 67–89, Cambridge University Press, Cambridge.

Lindsay, W.L., Hodgson, J.F. and Norvell, W.A. (1966) The physico-chemical equilibrium of metal chelates in soils and their influence on the availability of micronutrient cations: in *Soil Chemistry and Fertility, Transactions of Meeting of Commissions II and IV of the International Society of Soil Science*, 305–16, Aberdeen University Press, Aberdeen.

Linehan, D.J. (1978) Polycarboxylic acids extracted by water and by alkali from agricultural top soils of different drainage status. *Journal of Soil Science*, **29**, 373–7.

Linehan, D.J. (1985) Organic matter and trace elements in soil: in *Soil Organic Matter and Biological Activity*, Vaughan, D. and Malcolm, R.E., eds, 403–21, Nijhoff/Junk, Dordrecht, The Netherlands.

Lynch, J.M. (1980) Effects of organic acids on the germination of seeds and growth of seedlings. *Plant, Cell and Environment*, **3**, 255–9.

Lynch, J.M. (1983) *Soil Biotechnology*, Blackwell, Oxford.

Martell, A.E. (1975) The influence of natural and synthetic ligands on the transport and function of metal ions in the environment. *Pure and Applied Chemistry*, **44**, 81–113.

Martin, J.K. (1971) ^{14}C-labelled material leached from the rhizosphere of plants supplied with $^{14}CO_2$. *Australian Journal of Biological Science*, **24**, 1131–42.

McBride, M.B. (1978) Transition metal bonding in humic acid: An ESR study. *Soil Science*, **126**, 200–9.

Powlson, D.S. (1980) The effects of grinding on microbial and non-microbial organic matter in soil. *Journal of Soil Science*, **31**, 77–86.

Rovira, A.D. and Greacen, E.L. (1957) The effect of aggregate disruption on the activity of micro-organisms in the soil. *Australian Journal of Agricultural Research*, **8**, 659–73.

Russell, J.D., Vaughan, D., Jones, D. and Fraser, A.R. (1983) An IR spectroscopic study of soil humin and its relationship to other soil humic substances and fungal pigments. *Geoderma*, **29**, 1–12.

Schnitzer, M. and Kahn, S.U. (1978) *Soil Organic Matter*. Developments in Soil Science, Vol. 8. Elsevier, Oxford.

Skinner, F.A. (1975) Anaerobic bacteria and their activities in soil: in *Soil Microbiology: A Critical Review*, Walker, N., ed., pp. 1–19, Butterworth, London.

Stevenson, F.J. (1982) *Humus Chemistry – Genesis Composition, Reactions*, John Wiley & Sons, New York.

Strutt, N. (1970) *Modern Farming and the Soil*, Agricultural Advisory Council, HMSO, London.

Swift, M.J., Heal, O.W. and Anderson, J.M. (1979) *Decomposition in Terrestrial Ecosystems*. Studies in Ecology, Vol. 5, Blackwell, Oxford.

Swift, R.S. and Posner, A.M. (1972) Nitrogen, phosphorus and sulphur contents of

humic acids fractionated with respect to molecular weight. *Journal of Soil Science*, **23**, 50–7.
Tisdall, J.M. and Oades, J.M. (1982) Organic matter and water-stable aggregates in soils. *Journal of Soil Science*, **33**, 141–63.
Tisdale, S.L., Nelson, W.L. and Beaton, J.D. (1985) *Soil Fertility and Fertilizers*, fourth edition, Macmillan, New York.
Wagner, G.H. and Mutakar, V.K. (1968) Amino components of soil organic matter formed during humification of ^{14}C glucose. *Soil Science Society of America Proceedings*, **32**, 683–6.
Waksman, S.A. (1936) *Humus Origin, Chemical Composition and Importance in Nature*, Bailliere, Tindall and Cox, London.
Whittaker, R.H. (1975) *Communities and Ecosystems*, second edition, Macmillan, New York.

Chapter 4

Bowen, H.J.M. (1979) *Environmental Chemistry of the Elements*, Academic Press, London.
Bower, C.A. and Wilcox, L.V. (1965) Soluble salts: Chapter 62, in *Methods of Soil Analysis*, Part 2: *Chemical and Microbiological Properties*, Black, C.A., ed., American Society of Agronomy, Madison, Wisconsin.
Bowman, A.F., ed. (1990) *Soils and the Greenhouse Effect*. John Wiley and Sons, London.
Brady, N.C. (1990) *The Nature and Properties of Soils*, tenth edition, Macmillan, New York.
Brimblecombe, P. (1986) *Air Composition and Chemistry*, Cambridge University Press, Cambridge.
Cooke, G.W. (1967) *The Control of Soil Fertility*, Crosby Lockwood & Son, London.
Cresser, M.S. and Edwards, A.C. (1987) *Acidification of Freshwaters*, Cambridge University Press, Cambridge.
Follett, R.H., Murphy, L.S. and Donahue, R.A. (1981) *Fertilizers and Soil Amendments*, Prentice-Hall, New Jersey.
Freney, J.R. and Galbally, I.E. (1982) *Cycling of Carbon, Nitrogen, Sulfur and Phosphorus in Terrestrial and Aquatic Ecosystems*, Springer, Berlin.
Germida, J.J., Wainwright, M. and Gupta, V.V.S.R. (1992) Biogeochemistry of sulphur cycling in soil. *Soil Biochemistry*, **7**, 1–53.
Glinski, J. and Stepniewski, W. (1985) *Soil Aeration and its Role for Plants*, CRC Press, Florida.
Greenland, D.J. and Hayes, M.H.B. (1978) *The Chemistry of Soil Constituents*, John Wiley and Sons, Chichester.
Gupta, U.C. and Lipsett, J. (1981) Molybdenum in soils, plants and animals. *Advances in Agronomy*, **34**, 73–115.
Harrison, A.F., Ineson, P. and Heal, O.W. (1990) *Nutrient Cycling in Terrestrial Ecosystems*: *Field Methods, Application and Interpretation*, Elsevier, London.
Jacobs, C.W. (ed.) (1989) *Selenium in Agriculture and the Environment*, American Society of Agronomy, Madison, Wisconsin.
Jahiruddin, M., Chambers, B.J., Livesey, N.T. and Cresser, M.S. (1986) Effect of liming on extractable Zn, Cu, Fe and Mn in selected Scottish soils. *Journal of Soil Science*, **37**, 603–15.
Jeffrey, D.W. (1987) *Soil–Plant Relationships: An Ecological Approach*, Croom Helm, London.

References

Jones, U.S. (1982) *Fertilizers and Soil Fertility*, Reston Publishing Co., Virginia.

Landon, J.R. (1991) *Booker Tropical Soil Manual*, paperback edition, Longman, London.

Lindsay, W.L. (1979) *Chemical Equilibria in Soils*, John Wiley and Sons, New York.

Marr, I.L. and Cresser, M.S. (1983) *Environmental Chemical Analysis*, International Textbook Co., New York.

Ponnamperuma, F.N. (1985) Chemical kinetics of wetland rice soils relative to soil fertility. In *Wetland Soils: Characterization, Classification and Utilization*, pp. 71–89, International Rice Research Institute, Manila.

Rhoades, J.D. (1982a) Cation exchange capacity: Chapter 8 in *Methods of Soil Analysis, Part 2: Chemical and Microbiological Properties*, 2nd edition, Page, A.L., Miller, R.H. and Keeney, D.R., eds. American Society of Agronomy, Madison, Wisconsin.

Rhoades, J.D. (1982b) Soluble Salts: Chapter 10 in *Methods of Soil Analysis, Part 2: Chemical and Microbiological Properties*, 2nd edition, Page, A.L., Miller, R.H. and Keeney, D.R., eds. American Society of Agronomy, Madison, Wisconsin.

Russell, E.W. (1987) *Soil Conditions and Plant Growth*, eleventh edition, Longman, London.

Shorrocks, V.M. (1974) *Boron Deficiency – Its Prevention and Cure*, Borax Consolidated Ltd., Norwich.

Skiba, U. and Cresser, M.S. (1988) The ecological significance of increasing atmospheric carbon dioxide. *Endeavour*, **12**, 143–7.

Skiba, U., Edwards, A., Peirson-Smith, T. and Cresser, M. (1987) Rain simulation in acid rain research - Techniques, advantages and pitfalls. In *Chemical Analysis in Environmental Research*, Institute of Terrestrial Ecology, Merlewood.

Sposito, G. (1984) *The Surface Chemistry of Soils*, Oxford University Press, Oxford.

Thompson, L.M. and Troeh, F.R. (1978) *Soils and Soil Fertility*, McGraw Hill, New York.

Ure, A. (1991) Atomic absorption and flame emission spectroscopy: Chapter 1 in *Soil Analysis*, second edition, Smith, K.A. ed., Marcell Dekker, New York.

White, R.E. (1987) *Introduction to the Principles and Practice of Soil Science*, Blackwell Scientific Publications, Oxford.

Yu Tian-Ren (1985) *Physical Chemistry of Paddy Soils*, Springer-Verlag, Berlin.

Chapter 5

ADAS (1983) *Sampling Soil for Analysis*, MAFF leaflet 655.

AFRC Institute of Arable Crops Research (1989) *Annual Report*, Rothamsted.

Agricultural Lime Producers Council (1983) *Lime in Agriculture*, second edition, Claygate, Surrey.

Amer, F., Bouldin, D.R., Black, C.A. and Duke, F.R. (1955) Characterisation of soil phosphorus by anion exchange resin adsorption and P-32 equilibration. *Plant and Soil*, **6**, 391–408.

Anghinoni, I. and Barber, S.A. (1980) Predicting the most efficient phosphorus placement for corn. *Soil Science Society of America Journal*, **44**, 1016–20.

Atkinson, D. (1988) Action for analysis: Challenges for agricultural analysis. *Analytical Proceedings*, **24**, 118–20.

Austin, R.B., Bingham, J., Blackwell, R.D., Evans, L.T., Ford, M.M., Morgan, C.L. and Taylor, M. (1978) Genetic improvements in winter wheat yields since 1900

References

and associated physiological changes. *Journal of Agricultural Science (Cambridge)*, **94**, 657–89.

Bache, B.W. (1988) Measurement and mechanisms in acid soils. *Communications in Soil Science and Plant Analysis*, **19**, 775–92

Bache, B.W. and Crooke, W.M. (1981) Interactions between aluminium, phosphorus and pH in response of barley to soil acidity. *Plant and Soil*, **61**, 365–75.

Bache, B.W. and Ireland, C. (1980) Desorption of phosphate from soils using anion exchange resin. *Journal of Soil Science*, **31**, 297–306.

Bache, B.W. and Ross, J.A.M. (1991) Effect of phosphorus and aluminium in response of spring barley to soil acidity. *Journal of Agricultural Science, Cambridge*, **117**, 299–305.

Barber, S.A. (1984) *Soil Nutrient Bioavailability: Mechanistic Approach*, Wiley and Sons, New York.

Barrow, N.J. and Shaw, T.C. (1977) Factors affecting the amount of phosphate extracted from soil by anion exchange resin. *Geoderma*, **18**, 309–23.

Berrow, M.L., Burridge, J.C. and Reith, J.S.W. (1983) Soil drainage conditions and related plant trace element contents. *Journal of the Science of Food and Agriculture*, **34**, 53–4.

Berrow, M.L. and Mitchell, R.L. (1980) Location of trace elements in soil profiles: total and extractable contents of individual horizons. *Transactions of the Royal Society of Edinburgh: Earth Sciences*, **71**, 103–21.

Bibby, J.S., Douglas, H.A., Thomasson, A.J. and Robertson, J.S. (1982) *Land Capability Classification for Agriculture*. Soil Survey of Scotland Monograph. Macaulay Institute for Soil Research, Aberdeen.

Bolton, J. (1972) Changes in magnesium and calcium in soils of the Broadbalk wheat experiment of Rothamsted from 1865–1966. *Journal of Agricultural Science, Cambridge*, **79**, 217–23.

Brown, J.C. and Jones, W.F. (1977) Fitting plants nutritionally to soils. I Soyabeans. *Agronomy Journal*, **69**, 399–404.

Brown, S. and Lugo, A.E. (1990) Effects of forest clearing and succession on the carbon and nitrogen content of soils in Puerto Rico and U.S. Virgin Islands. *Plant and Soil*, **124**, 53–64.

Carter, D.L. (1982) Salinity and plant productivity. In *Handbook of Agricultural Productivity*, Volume 1, Rechcigl, M. Jr., ed., CRC Press, Florida.

Church, B.M. (1981) Use of fertilizers in England and Wales. *Report of Rothamsted Experimental Station for* 1980, part 2, pp 115–22.

Church, B.M. and Skinner, R.J. (1986) The pH and nutrient status of agricultural soils in England and Wales 1969–83. *Journal of Agricultural Science, Cambridge*, **107**, 21–8.

Cipra, J.E., Bidwell, O.W., Whitney, D.A. and Feyerherm, A.M. (1972) Variations with distance in selected fertility measurements of pedons of Western Kansas Ustoll. *Soil Science Society of America Proceedings*, **36**, 111–5.

Cope, J.T. and Evans, C.E. (1985) Soil testing. *Advances in Soil Science*, **1**, 201–28.

Cottenie, A. (1980) Soil and plant testing as a basis of fertiliser recommendations. *FAO Soils Bulletin* 38/2.

Couto, W. (1982) Soil pH and plant productivity: in *Handbook of Agricultural Productivity*, Volume 1, Rechcigl, M. Jr., ed., CRC Press, Florida.

Curtin, D. and Smillie, G.W. (1986) Effects of liming on soil chemical characteristics and grass growth in laboratory and long-term field-amended soils. *Plant and Soil*, **95**, 15–22.

Dickson, E.L. and Stevens, R.J. (1983) The influence of parent material and fertility status on certain extractable trace elements in Northern Ireland soils. *Journal of the Science of Food and Agriculture*, **34**, 52–3.

ECETOC, (1988) *Nitrate in Drinking Water: Technical Report* 27 European Chemical Industry Ecology and Toxicology Centre, Brussels, Belgium.

Farr, E., Vaidynathan, L.V. and Nye, P.H. (1969) Measurement of ionic concentration gradients in soil near roots. *Soil Science*, **107**, 385–91.

Fertilizer Product Consumption (1989). International Fertilizer Industry Association Limited, Paris.

Fox, R.H. and Hoffman, L.D. (1981) The effect of N fertilizer source on grain yield, nitrogen uptake, soil pH, and lime requirement in no-till corn. *Agronomy Journal*, **73**, 891–5.

Foy, C.D. (1988) Plant adaption to acid, and aluminum toxic soils. *Communications in Soil Science and Plant Analysis*, **19**, 959–88.

Foy, C.D., Chaney, R.L. and White, M.C. (1978) The physiology of metal toxicity in plants. *Annual Reviews of Plant Physiology*, **29**, 511–66.

Gasser, J.K.R. (1973) An assessment of the importance of some factors causing losses of lime from agricultural soils. *Experimental Husbandry*, **25**, 86–95.

Glentworth. R. and Muir, J.W. (1963) *The Soils Round Aberdeen, Inverurie and Fraserburgh*, Soil Survey of Scotland, Macaulay Institute for Soil Research, Aberdeen.

Goss, M.J., Williams, B.L. and Howse, K.R. (1991) Organic matter turnover and nitrate leaching. In *Advances in Soil Organic Matter Research: The Impact of Agriculture and the Environment*, Wilson, W.S., ed, Special Publication No. 90, Royal Society of Chemistry, Cambridge.

Goulding, K.W.T., Powlson, D.W., Poulton, P.R., Webster, C.P., Macdonald, A.J. and Glending, M.J. (1989) Implications to nitrogen losses of a switch from inorganic towards organic manures and from continuous arable cropping to ley-arable rotations. *Journal of the Science of Food and Agriculture*, **48**, 123–4.

Hanotiaux, G. and Vanoverstraeten, M. (1990) Study of the utilisation of mineral phosphate fertilisers in Western Europe. *Scientific Report, IMPHOS, Maroc*.

Haynes, R.J. (1982) Effect of liming on phosphate availability in acid soils. A critical review. *Plant and Soil*, **68**, 289–308.

Haynes, R.J. (1990) Movement and transformation of fertiliser nitrogen below trickle emitters and their effects on pH in the wetted soil volume. *Fertilizer Research*, **23**, 105–12.

Hornung, M., Stevens, P.A. and Reynolds, B. (1986) The impact of pasture improvement on the soil solution chemistry of some stagnopodzols in Mid-Wales. *Soil Use and Management*, **2**, 18–26.

Jackson, R.B., Manwaring, J.H. and Caldwell, M.M. (1990) Rapid physiological adjustment of roots to localised soil enrichment. *Nature (London)*, **344**, 58–60.

Jacobsen, J.S. and Westerman, R.L. (1991) Stratification of soil acidity derived from nitrogen fertilization in winter wheat tillage systems. *Communications in Soil Science and Plant Analysis*, **22**, 1335–46.

Kang, B.T., Wilson, G.F. and Spiken, L. (1981) Alley cropping maize (*Zea Mays L.*) and Lelucaena (*Leucaena Laucocephala LAM*) in southern Nigeria. *Plant and Soil*, **63**, 165–79.

Kuhlmann, H. and Baumgartel, G. (1991) Potential importance of the subsoil for the phosphorus and magnesium nutrition of wheat. *Plant and Soil*, **137**, 259–66.

References

Lal, R. (1986) Conversion of tropical rainforest: agronomic potential and ecological consequences. *Advances in Agronomy*, **39**, 173–264

Lal, R. and Ghaman, B.S. (1989) Biomass burning in windrows after clearing a tropical rain forest: Effect on soil properties, evaporation and crop yields. *Field Crops Research*, **22**, 247–55.

Larsen, S. (1967) Soil phosphorus. *Advances in Agronomy*, **19**, 151–210.

Lee, L.K. and Neilsen, E.G. (1987) The extent and costs of groundwater contamination by agriculture. *Journal of Soil and Water Conservation*, July-August, 243–8.

Macaulay Institute for Soil Research (1984) *Annual Report No.* 54, Macaulay Institute for Soil Research, Aberdeen.

Macaulay Institute for Soil Research (1986) *Annual Report No.* 56, Macaulay Institute for Soil Research, Aberdeen.

MacDonald, K.B. and Brklacich, M. (1992) Prototype agricultural land evaluation systems for Canada: 1 Overview of systems development. *Soil Use and Management*, **8**, 1–8.

Madison, R.J. and Brunett, J.O. (1984) Overview of the occurrence of nitrate in groundwater of the United States. *US Geological Survey Water Supply Paper*, 2275, 93–105.

Marschner, H. (1991) Plant-soil relationships: acquisition of mineral nutrients by roots from soil: in *Plant Growth Interactions with Nutrition and Environment*, Porter, J.R. and Lawlor, D.W., eds, Seminar Series 43, Society of Experimental Biologists.

McRae, R.J., Hill, S.B., Mehys, G.R. and Henning, J. (1990) Farm-scale agronomic and economic conversion from conventional to sustainable agriculture. *Advances in Agronomy*, **43**, 155–98.

Mengel, K. (1982) Factors of plant nutrient availability relevant to soil testing. *Plant and Soil*, **64**, 129–38.

Millard, P. and Neilsen, G.H. (1989) The influence of nitrogen supply on the uptake and remobilization of stored nitrogen for the seasonal growth of apple trees, *Annals of Botany*, **63**, 301–9.

Moorby, H., White, R.E. and Nye, P.H. (1988) The influence of phosphorus nutrition and H ion efflux from the roots of young rape plants. *Plant and Soil*, **105**, 247–56.

Neilsen, G.H. and Hoyt, P.B. (1982) Soil pH variation in British Columbia apple orchards. *Canadian Journal of Soil Science*, **62**, 695–8.

Neilsen, G.H. Hoque, E. and Drought, B.G. (1981) The effects of surface-applied calcium on soil and mature Spartan apple trees. *Canadian Journal of Soil Science*, **61**, 295–302.

Nye, P.H. (1981) Changes in pH across the rhizosphere induced by roots. *Plant and Soil*, **61**, 7–26.

Nye, P.H. and Tinker, P.B. (1977) *Solute movement in the soil-root system*. Studies in Ecology, Vol. 4, Blackwell, Oxford.

Olsen, S.R. and Khasawneh, F.E. (1980) Use and limitations of physical-chemical criteria for assessing the status of phosphorus in soils: in *The Role of Phosphorus in Agriculture*, Khasawneh, F.E., Sample, E,C. and Kamprath, E.J., eds, American Society of Agronomy, Crop Science Society of America and Soil Science Society of America, Madison, Wisconsin.

Porter, J.R. and Lawlor, D.W. eds. (1991) *Plant Growth Interactions with Nutrition and Environment*, Society for Experimental Biology Seminar Series 43, Cambridge University Press, Cambridge.

Proctor, J. (1987) Nutrient cycling in primary and old secondary rainforests. *Applied Geography*, **7**, 135–52.

Reith, J.W.S. (1980) Trends in the nutrient status of Scottish soils. *Proceedings of the Third Study Conference of the Scottish Agricultural Colleges*, Ayr, UK, Scottish Agricultural College, Edinburgh.

Reith, J.W.S., Inkson, R.H.E., Caldwell, K.S., Simpson, W.E. and Ross, J.A.M. (1984) Effect of soil type and depth on crop production. *Journal of Agricultural Science, Cambridge*, **103**, 377–86.

Riggs, T.J., Hanson, P.R., Start, N.D., Miles, D.M., Morgan, C.L. and Ford, M.A. (1981) Comparison of spring barley varieties grown in England and Wales between 1880–1980. *Journal of Agricultural Science, Cambridge*, **97**, 599–610.

Rosswall, T. and Paustian, K. (1984) Cycling of nitrogen in modern agricultural systems. *Plant and Soil*, **76**, 3–21.

Royal Society (London) (1983) *The Nitrogen Cycle of the United Kingdom. A Study Group Report*. The Royal Society, London.

SAC/MISR (1985) *Fertiliser Recommendations*, Macaulay Institute for Soil Research, Aberdeen.

Sadusky, M.C., Sparks, D.L., Noll, M.R. and Hendricks, G.J. (1987) Kinetics and mechanisms of potassium release from sandy Middle Atlantic Coastal Plain soils. *Soil Science Society of America Journal*, **51**, 1460–5.

Sanyal, S.K. and De Datta, S.K. (1991) Chemistry of P transformations in soil. *Advances in Soil Science*, **16**, 1–120.

Skinner, R.J., Church, B.M. and Kershaw, C.D. (1992) Recent trends in soil pH and nutrient status in England and Wales. *Soil Use and Management*, **8**, 16–20.

Smith, K.A., Elmes, A.E., Howard, R.S. and Franklin, M.F. (1984) The uptake of soil and fertiliser nitrogen by barley growing under Scottish climatic conditions. In *Biological Processes and Soil Fertility*, Tinsley, J. and Darbyshire, J.F., eds, Developments in Plant and Soil Sciences, II, Junk, The Netherlands.

Smith, S.S. and Young, L.B. (1975) Distribution of N forms in virgin and cultivated soils. *Soil Science*, **120**, 354–60.

Somasiri, L.L.W. and Edwards, A.C. (1992) An ion exchange resin method for nutrient extraction of agricultural advisory soil samples. *Communications in Soil Science and Plant Analysis*, **23**, 645–57.

Sparks, D.L. and Huang, P.M. (1985) Physical chemistry of soil potassium: in *Potassium in Agriculture*. Munson, R.D., ed., American Society of Agronomy, Crop Science Society of America and Soil Science Society of America, Madison, Wisconsin.

Sumner, M.E. and Carter, E. (1988) Amelioration of subsoil acidity. *Communications in Soil Science and Plant Analysis*, **19**, 1309–18.

Taylor, G.J. (ed.) (1988) Plant–soil interactions at low pH. Special Issue *Communications in Soil Science and Plant Analysis*, **19**.

Walker, J.M. and Barber, S.A. (1961) Ion uptake by living plant roots. *Science*, **133**, 881–2.

White, G.C. (1961) A survey of the pH of the soils of East Malling Research Station. In *East Malling Research Station Annual Report*, East Malling Research Station.

Chapter 6

Billett, M.F. and Cresser, M.S. (1992) Predicting stream water quality using catchment and soil chemical characteristics. *Environmental Pollution*, **77**, 263–8.

References

Bishop, K.H., Grip, H. and O'Neill, A. (1990). The origins of acid runoff in a hillslope during storm events. *Journal of Hydrology*, **116**, 35–61.

Boesten, J.J.T.I. (1987) Leaching of herbicides to groundwater: A review of important factors and of available measurements: in *British Crop Protection Conference Proceedings, Volume 2, Weeds*, 559–68. British Crop Protection Council, Thornton Heath, Surrey.

Christophersen, N., Neal, C., Hooper, R.P., Vogt, R.D. and Andersen, S. (1990a) Modelling streamwater chemistry as a mixture of soilwater end-members – A step towards second generation acidification models. *Journal of Hydrology*, **116**, 307–20.

Christophersen, N., Robson, A., Neal, C., Whitehead, P.G., Vigerust, B. and Henriksen, A. (1990b) Evidence for long-term deterioration of streamwater chemistry and soil acidification at the Birkenes catchment, Southern Norway. *Journal of Hydrology*, **116**, 63–76.

Crawshaw, D.H. (1986) The effects of acidic runoff on streams in Cumbria. In *Pollution in Cumbria*, 25–32. Institute of Terrestrial Ecology, Huntingdon.

Cresser, M. and Edwards, A. (1987) *Acidification of Freshwaters*, Cambridge University Press, Cambridge.

de Vries, W., Posch, M. and Kämäri, J. (1989) Simulation of the long-term soil response to acid deposition in various buffer ranges. *Water, Air, and Soil Pollution*, **48**, 349–90.

Driscoll, C.T., Johnson, N.M., Likens, G.E. and Feller, M.C. (1988) Effects of acid deposition on the chemistry of headwater streams: A comparison between Hubbard Brook, New Hampshire, and Jamieson Creek, British Columbia. *Water Resources Research*, **24**, 195–200.

Dunning, J.C. (1986) The farmer and pollution: in *Pollution in Cumbria*, pp. 33–8, Institute of Terrestrial Ecology, Huntingdon.

Edwards, A.C., Creasey, J. and Cresser, M.S. (1986) Soil freezing effects on upland stream solute chemistry. *Water Research*, **20**, 831–4.

Elinder, C.G. (1984) Metabolism and toxicity of metals: in *Changing Metal Cycles and Human Health*, pp. 265–84, Nriagu, J.O., ed., Springer-Verlag, Berlin.

Harvey, H.H. (1985) The biological evidence of acidification, mechanism of action and an attempt at predicting acidification effects: in *Symposium on the Effects of Air Pollution on Forest and Water Ecosystems, Helsinki, April 23–24, 1985*, pp. 63–78. Foundation for Research of Natural Resources in Finland, Helsinki.

Higgins, I.J. and Burns, R.G. (1975) *The Chemistry and Microbiology of Pollution*, Academic Press, London.

Johnson, D.W., Cresser, M.S., Nilsson, S.I., Turner, J., Ulrich, B., Binkley, D. and Cole, D.W. (1991) Soil changes in forest ecosystems: evidence for and probable causes. *Proceedings of the Royal Society of Edinburgh*, **97B**, 81–116.

Joyce, D.A. (1986) Agricultural pollution: The role of the Agricultural Development and Advisory Service in pollution problems: in *Pollution in Cumbria*, pp. 32–5, Institute of Terrestrial Ecology, Huntingdon.

Krug, E.C. and Frink, C.R. (1983) Acid rain on acid soil: A new perspective. *Science*, **221**, 520–5.

Lelong, F., Dupraz, C., Durand, P. and Didon-Lescot, J.F. (1990) Effects of vegetation on the biogeochemistry of small catchments (Mont Lozere, France). *Journal of Hydrology*, **116**, 125–45.

Mason, B.J. (ed.) (1991) *The Surface Waters Acidification Programme*, Cambridge University Press, Cambridge.

Sanger, L., Billett, M.F. and Cresser, M.S. (1992) Assessment by laboratory simulation of approaches to amelioration of peat acidification. *Environmental Pollution*, in press.

Stone, A., Seip, H.M., Tuck, S., Jenkins, A., Ferrier, R.C. and Harriman, R. (1990) Simulation of hydrochemistry in a highland Scottish catchment using the Birkenes model. *Water, Air, and Soil Pollution*, **51**, 239–59.

Walling, D.E. (1980) Water in the catchment ecosystem: in *Water Quality in Catchment Ecosystems*, pp. 1–47, Gower, A.M., ed., John Wiley & Sons, Chichester.

Walling, D.E. and Webb, B.W. (1981) Water quality: in *British Rivers*, pp. 126–69. Lewin, J., ed., George Allen & Unwin, London.

Water Act (1989) Schedule 2 of the Water Supply (Water Quality) Regulations, HMSO, London.

Whitehead, P.G. (1990) Modelling nitrate from agriculture into public water supplies. *Philosophical Transactions of the Royal Society*, London, **B329**, 403–10.

Wolock, D.M., Hornberger, G.M. and Musgrove, T.J. (1990) Topographic effects on flow path and surface water chemistry of the Llyn Brianne catchments in Wales. *Journal of Hydrology*, **115**, 243–59.

Wright, R.F., Norton, S.A., Brakke, D.F. and Frogner, T. (1988) Experimental verification of episodic acidification of freshwaters by seasalts. *Nature (London)*, **334**, 422–4.

Chapter 7

Billett, M.F., Fitzpatrick, E.A. and Cresser, M.S. (1988) Long-term changes in the acidity of forest soils in north-east Scotland. *Soil Use and Management*, **4**, 102–7.

Billett, M.F., Parker,-Jervis, F., Fitzpatrick, E.A. and Cresser, M.S. (1990) Forest soil chemical changes between 1949/50 and 1987. *Journal of Soil Science*, **41**, 133–45.

Billett, M.F., Fitzpatrick, E.A. and Cresser, M.S. (1991) Long-term changes in the Cu, Pb and Zn content of forest soil organic horizons from north-east Scotland. *Water, Air, and Soil Pollution*, **59**, 179–91.

Bridges, E.M. (1991) Dealing with contaminated soil. *Soil Use and Management*, **7**, 151–8.

Bull, K., Hall, J., Steenson, D., Smith, C. and Cresser, M.S. (1992) Critical loads of acid deposition for soils: The UK approach. *Endeavour*, **16**, 132–8.

Capriel, H., Haisch, A. and Khan, S.U. (1985) Distribution and nature of bound (non-extractable) residues of atrazine in a mineral soil nine years after the herbicide application. *Journal of Agricultural and Food Chemistry*, **33**, 567–9.

Crathorne, B. and Dobbs, A.J. (1990) Chemical pollution of the aquatic environment by priority pollutants and its control: Chapter 1 in *Pollution Causes, Effects and Control*, pp. 1–18, Harrison, R.M., ed., Royal Society of Chemistry, Cambridge.

Dios Cancela, G., Romero Taboada, E. and Sanchez-Rasero, F. (1992) Carbendazim adsorption on montmorillonite, peat and soils. *Journal of Soil Science*, **43**, 99–111.

Driscoll, C.T., Johnson, N.M., Likens, G.E. and Feller, M.C. (1988) Effects of acidic deposition on the chemistry of headwater streams: A comparison between Hubbard Brook, New Hampshire, and Jamieson Creek, British Columbia. *Water Resources Research*, **24**, 195–200.

Dunigan, E.P. and McIntosh, T.H. (1971) Atrazine–soil organic matter interactions. *Weed Science*, **19**, 279–82.

El-Sayad, E.A. (1988) *Status of Some Trace Elements in Relation to the Nature of the Main Sediments in the Fayoum (Egypt) Depression*, PhD Thesis, University of Aberdeen.

Edwards, A.C., Pugh, K., Wright, G., Sinclair, A.H. and Reaves, G.A. (1990) Nitrate status of two major rivers in NE Scotland with respect to land use and fertiliser additions. *Chemistry and Ecology*, **4**, 97-107.

Johnson, A.H., Siccama, T.G., Silver, W.L. and Battles, J.J. (1989) Decline of red spruce in high elevation forests of New York and New England. In *Acidic Precipitation. Volume 1: Case Studies*, pp. 85–112, Adriano, D.C. and Havas, M., eds, Springer-Verlag, New York.

Johnson, D.W. and Lindberg (1989) Acidic deposition on Walker Branch watershed. In *Acidic Precipitation. Volume 1: Case Studies*, pp. 1–38, Adriano, D.C. and Havas, M., eds, Springer-Verlag, New York.

Lester, J.N. (1990) Sewage and sewage sludge treatment: Chapter 3 in *Pollution Causes, Effects and Control*, pp. 33–62, Harrison, R.M., ed., Royal Society of Chemistry, Cambridge.

Li, G.C. and Felbeck, Jr., G.T. (1972) A study of the mechanism of atrazine adsorption by humic acid from muck soil. *Soil Science*, **113**, 140–8.

Loehr, R.C., Martin, J.H. and Neuhauser, E.F. (1992) Land treatment of an aged oily sludge – Organic loss and change in soil characteristics. *Water Research*, **26**, 805–15.

Malanchuk, J.L. and Nilsson, J., eds (1989) *The Role of Nitrogen in the Acidification of Soils and Surface Waters*, Nordic Council of Ministers, Copenhagen, Denmark.

Mitchell, G. and McDonald, A.T. (1992) Discolouration of water by peat following induced drought and rainfall simulation. *Water Research*, **26**, 321–6.

Nilsson, J. (ed.) (1986) *Critical Loads for Sulphur and Nitrogen, Report from a Nordic Working Group*, The Nordic Council of Ministers, Stockholm.

Posch, M., Falkengren-Grerup, U. and Kauppi, P. (1989) Application of two soil acidification models to historical soil chemistry data from Sweden: Chapter 18 in *Regional Acidification Models: Geographic Extent and Time Development*, pp. 241–51, Kamari, J., Brakke, D.F., Jenkins, A., Norton, S.A. and Wright, G.F., eds, Springer-Verlag, Berlin.

Quine, T.A. and Walling, D.E. (1991) Rates of soil erosion on arable fields in Britain: Quantitative data from caesium–137 measurements. *Soil Use and Management*, **7**, 169–76.

Reuberg, I., Brodin, Y.W., Cronberg, G., El-Daoushy, F., Oldfield, F., Rippey, B., Sandoy, S., Wallin, J.E., and Wilk, M. (1990) Recent acidification and biological changes in Lilla Oresjon, south west Sweden, and the relation to atmospheric pollution and land use history. *Philosophical Transactions of the Royal Society, London*, **327B**, 391–6.

Rippey, B. (1990) Sediment chemistry and atmospheric contamination. *Philosophical Transactions of the Royal Society, London*, **327B**, 311–7.

Sauerbeck, D. (1987) Effects of agricultural practices on the physical, chemical and biological properties of soils: Part II – Use of sewage sludge and agricultural wastes. In *Scientific Basis for Soil Protection in the European Community*, Barth, H. and Hermite, P.L., eds, Elsevier, London.

Skiba, U. and Cresser, M.S. (1989) Prediction of long- term effects of rainwater acidity on peat and associated drainage water chemistry in upland areas. *Water Research*, **23**, 1477–82.

Skiba, U., Cresser, M.S., Derwent, R.G. and Futty, D.W. (1989) Peat acidification in Scotland. *Nature (London)*, **337**, 68–9.

Stevens, P.A. and Wannop, C.P. (1987) Dissolved organic nitrogen and nitrate in an acid forest soil. *Plant and Soil*, **102**, 137–9.

Tamura, T. and Jacobs, D.G. (1960) Structural implications in caesium sorption. *Health Physics*, **2**, 391–8.

van Breemen, N., Boderie, P.M.A. and Booltink, H.W.G. (1989) Influence of airborne ammonium sulphate on soils of an oak woodland ecosystem in the Netherlands: Seasonal dynamics of solute fluxes: in *Acid Precipitation*, Volume 1: *Case Studies*, Adriano, D.C. and Havas, M., eds, Springer-Verlag, New York.

Vink, J.P.M. and Robert, P.C. (1992) Adsorption and leaching behaviour of the herbicide alachlor (2-chloro–2',6'diethyl-N-(methoxymethyl) acetanilide) in a soil specific management. *Soil Use and Management*, **8**, 26–30.

Walling, D.E. and Quine, T.A. (1991) Use of ^{137}Cs measurements on arable fields in the UK: potential applications and limitations. *Journal of Soil Science*, **42**, 147–65.

Weber, J.B., Weed, S.B. and Ward, T.M. (1969) Adsorption of s-triazines by soil organic matter. *Weed Science*, **17**, 417–21.

Wellburn, A. (1988) *Air Pollution and Acid Rain: The Biological Impact*. Longman Scientific and Technical, Harlow, England.

Witter, E. (1991) Use of $CaCl_2$ to decrease ammonia volatilization after application of fresh and anaerobic chicken slurry to soil. *Journal of Soil Science*, **42**, 369–80.

Woodward, F.I. (1988) The response of stomata to changes in atmospherical levels of CO_2. *Plants Today*, **1**, 132–5.

Chapter 8

Aitken, R.L., Moody, P.W. and Compton, B.L. (1990) A simple bioassay for the diagnosis of aluminium toxicity in soils. *Communications in Soil Science and Plant Analysis*, **21**, 511–29.

El-Sayad, E., Cresser, M.S., El-Gawad, M.A. and Khater, E.A. (1988) The determination of carbonate fraction trace metals in calcareous soils. *Microchemical Journal*, **38**, 307–12.

Harrison, A.F. and Dighton, J. (1990) Determination of phosphorus status of wheat and barley crops using a rapid root bioassay. *Journal of the Science of Food and Agriculture*, **51**, 171–7.

Jarvis, K.E., Gray, A.L. and Houk, R.S. (1991) *Handbook of Inductively Coupled Plasma Mass Spectrometry*, Blackie, Glasgow.

Potts, P.J. (1987) *A Handbook of Silicate Rock Analysis*, Blackie, Glasgow.

Schulthess, C.P. and Sparks, D.L. (1991) Equilibrium-based modelling of chemical sorption on soils and soil constituents. *Advances in Soil Science*, **16**, 121–63.

Tessier, A. and Campbell, P.G.C. (1991) Comment on 'Pitfalls of Sequential Extractions' by P.M.V. Nirel and F.M.M. Morel, Water Research, 24, 1055–6 (1990); *Water Research*, **25**, 115–17.

INDEX

Aberdeen Soil Horizon Model, 143–5
acetic acid, 31, 53–4
acid rocks, 13
acidification
   modelling, 146, 159–62
   soil, 3, 12, 157–62
acidity, 113–23
Acromobacter, 51
actinomycetes, 33
aeration, 49–54, 83–90
albite, 19, 21
algae, 8, 45
alkanes, 41, 153
allophane, 19
aluminium
   coordination of, 22
   effects on health, 129
   in rocks, 11
   in waters, 128–9, 138, 147, 158
   substitution for, 23, 59
   substitution of, 23
   toxicity, 116, 117, 119, 172
amino acids, 41
ammonia, 92
   as a pollutant, 155, 157
   in soil atmosphere, 7
   in water, 129
   oxidation, 78, 157
   volatilisation, 109
amphibole, 20
amphibolite, 12
anaerobiosis, 3, 49–54, 85–90
analysis
   advisory, 123
   foliar, 123
   soil, 124–7
anatase, 19
andesite, 12, 13
anhydrite, 16

animal wastes
   collection of, 98
   see also manure
anion exchange, 73–4, 158
anorthite, 19, 22
antigorite, 19
apatite, 19, 24
aragonite, 19
archaeology, 1, 5–6
arkrose, 17
arsenic, 79
Arthrobacter, 33
Aspergillus, 31
atmosphere
   pollution of, 93–5, 155–7, 162–3
   regulation of, 1
   soil, 7, 134–6
atomic absorption, 60, 78, 171
augite, 11, 12, 15, 20
autochthonous decomposers, 36–7

Bacillus, 31, 51
bacteria, 31, 36
basalt, 11, 12, 13
base cations, 3, 11, 16
basic rocks, 12, 13, 26
beidelite, 19
bioassays, 172
biotite, 11, 12, 15, 19, 24
Birkenes Model, 146
boehmite, 19
boron
   cycling of, 91
   deficiency of, 54, 77
   effect of pH on, 77
   essential element, 2
   forms of, 69, 77
   in minerals, 54
   toxicity, 77

Index

brickearth, 17
bromine, 79
brown rot fungi, 32
brucite, 22
butyric acid, 54

cadmium
  as a pollutant, 152–3
  in fertiliser, 153
  in sludge, 152
caesium
  as a pollutant, 155–6
  mobility of, 156
calcite, 19
calcium
  as a bridge, 40
  carbonate, 14, 16, 19, 60, 79–80, 125, 157
  cycling of, 68
  effect of pH on, 69–70, 116–17
  essential element, 2, 8
  in freshwaters, 129, 158
  outputs in rivers, 18, 142
  saturation, 145
carbon
  cycling, 3, 28, 35, 92–3, 162–3
  global reservoir, 28
carbon dioxide, 7, 30, 35
  effects on water pH, 134–6
  increases in atmospheric, 162–3
  reduction of, 51
carbonate, 14, 16
cation exchange, 25, 55, 58–62, 131, 138, 145, 149, 153–4, 157
cation-exchange capacity, 55, 145
  measurement of, 60–1
cellobiase, 32
cellotriase, 32
cellulase, 31, 32
cellulose, 31, 38, 53
charge transfer bonding, 164
chelation, 56, 164–5
chemisorption, 165
Chernobyl, 155–6
chert, 16
chlorine, cycling of, 95
chlorite, 19
chromite, 11
chromium
  as a pollutant, 152
  essential element, 79
clay minerals, 24–7
climate effects
  arid, 3, 79–83, 148
  humid, 132
Clostridium, 51

coal, 16
coatings, 14
cobalt
  cycling of, 91
  effect of pH on, 70–1
  effect of soil drainage on, 102
  essential element, 2
  in rocks, 11
colour
  of minerals, 15, 24, 26
  significance of, 57
conchoidal fracture, 12, 15
conductance, 81
conductivity, 81
conglomerate, 17
coniferyl alcohol, degradation of, 34
copper
  as a pollutant, 152
  cycling of, 91
  deficiency of, 54
  effect of drainage, 102
  effect of pH on, 70–1
  essential element 2
  forms of, 69
  in rocks, 11
  species of, 55
core, 10
Coriolus, 35
corundum, 16
critical load, 5, 162
crop rotation, 103–5
crust, 11
crystobalite, 19
cycling
  methods for studying, 95–9
  nutrient, 3, 35, 39, 90–9

dating, radiocarbon, 38
DDT, 54
denitrification, 51, 90, 93, 94, 109, 148
depolymerisation, 31–2, 35
depth of soil, effect on fertility, 102
Desulfovibrio, 51
diagenesis, 16
diffusion coefficients, 125
dioctahedral sheet, 22
diopside, 19, 20
diorite, 12, 13
dolerite, 12, 13
dolomite, 19
  metamorphism of, 23
drainage effects
  on biological activity, 3
  on water quality, 147–8
  see also anaerobiosis
drainage water, sampling of, 98

Index

drinking water, 128, 154, 163
dry deposition
  measurement of, 97
  of pollutants, 155

earthworms, 36
ecosystem approach, 169–70
Eh, 84–8
electron spin resonance, 56
endo-enzymes, 33
endopolymerase, 31
enstatite, 11, 20
enzymes, 31–3
erosion, 47, 147
essential elements, 2
  forms of, 69
ethane, 92
ethanol, 53
ethylene, 92
evolution, soil, 4, 8, 141, 143, 168
exo-enzymes, 33
exopolymerase, 31
extractants in trace analysis, 127
exudates, root, 3, 29

facultative anaerobes, 51
fallow, 104
fatty acids, 41
feldspars, 11, 21, 23
fermentation, 53, 90
fertiliser
  assessment of requirement, 123
  compound, 107
  effects on water quality, 149–50, 154
  fate of, 4, 109
  pollutants in, 152, 153–4
  use of, 106–13
fertility, 1
  definition, 100
  effect of parent material, 101
  evaluation of, 123–7
  fertiliser effects on, 106–13
  harvest effects on, 101
  management effects on, 103–6
  sustainable, 100–5
Fick's Law, 52
flint, 16
fluorine, 79
food chain, 1, 5, 163
formation, *see* evolution
freezing effects, 140
freshwater, quality of, *see* water
Fuller's earth, 17
fulvic acid, 44, 56
fungi, 8, 31, 33, 36, 45

fungicide, 2
Fusarium, 31

gabbro, 12, 13, 101
galacturonic acid, 33
ganister, 17
Gapon's Equation, 62
genetically engineered organisms, 169
geography, 1
gibbsite, 19, 22, 23
glauconite, 19
gleying, 102, 144
glucose, 32, 38, 53
goethite, 19
granite, 12, 13, 101, 141, 146, 158
greywacke, 17
grinding, effects of, 42
groundwater, pollution of, 110
gypsum, 14, 16, 19, 60, 80, 83

halite, 19
halloysite, 19
heavy metals, 152, 155
hematite, 16, 19
hemicellulose, 32–3
herbicide, 2
heterogeneity, soil, 103, 122
heterotrophic organisms, 32
history, 1, 6–7
hornblende, 12, 15, 19
Hubbard Brook, 158
humic acid, 43–4, 56
humic material, 30, 43
  clay complexes of, 40
  fractionation of, 43
  pesticide adsorption by, 166
humin, 44, 56
hydrogen
  cycling of, 91–2
  reduction to, 84
hydrogen bonding, 25, 164
hydrogen sulphide, 7, 51, 90, 92
hydrological cycle, 91–2, 130
hydrological pathway, 4, 128, 130–4, 144
hydrophobic bonding, 164
hydrous micas, 25
hydrous oxides, 73, 74, 78, 158–9
hypersthene, 20
hypomagnesaemia, 113

ice core analysis, 155
igneous rocks, 11–14
ilmenite, 11, 12
inductively coupled plasma, 171
inner-sphere complex, 59
intercropping, 105

Index

intermediate rocks, 13
interstratification, 19
ion exchange
   in pesticide adsorption, 164
   see cation exchange, anion exchange
iron,
   as substitute for aluminium, 23
   cycling of, 91
   deficiency of, 54
   effect of pH on, 72–3
   essential element, 2
   oxidation of, 140, 148
   reduction of, 51, 84, 85, 89
ironstone, 16
isomorphous substitution, 21–2, 23, 58, 59
isotopic exchange, 82, 98

jasper, 16

kaolinite, 19, 24–5
kinetics of organic matter breakdown, 37–8

labile elements, 124
land evaluation, 102
lattice energy, 18, 23
leaching
   foliar, 3, 68, 94
   nitrogen losses, 109–10
lead
   as a pollutant, 4, 152, 155
   in acid waters, 128
lichen, 8
ligand exchange, 164
lignin, 32, 38
   pesticide adsorption by, 166
   structure of, 34–5
lignite, 16
lime
   losses of, 120–1
   penetration of effect, 122, 142
limenite, 19
liming, 119–22
   effects on water quality, 150
limonite, 19
Lindane, 54
lipids, 53
litter, 68, 94, 96
   collection of, 97–8
   degradation, 3, 28, 98
   load, 5
loess, 17

magma, 11

magnesium
   as substitute for aluminium, 23
   coordination of, 22
   cycling of, 68
   effect of pH on, 69–70
   essential element, 2, 8
   imbalance, 113
   in rocks, 11
   outputs in rivers, 18, 142
magnetite, 11, 12
manganese
   cycling of, 91
   deficiency of, 54, 107
   effect of pH on, 72–3
   essential element, 2
   forms of, 51, 69
   oxidation of, 148
   reduction of, 51, 84, 85, 90
mantle, 11
manure, 4, 103, 149, 152–3, 157
mercury as a pollutant, 152
metamorphism, 23
Methanobacterium, 51
Michaelis–Menten kinetics, 36–7
microbial biomass, 29, 42, 105
microcline, 19
microdiorite, 13
microgranites, 12, 13
microorganisms
   activity of, 3, 31–7, 163
   and organic matter turnover, 35–8
   as organic matter, 28
   as pollutants, 152
mineralisation, *see* nitrogen mineralisation
minerals
   1:1 type, 23–4, 25
   2:1 type, 23–4, 25
   clay, 24–7
   importance, 3
   primary, 14, 24
   secondary, 14, 18
   see also under individual mineral names
mites, 36
modelling
   approaches to, 169
   organic matter turnover, 47–9
   soil acidification, 146, 161
   water quality, 143–7
molluscicide, 2
molybdenum
   anion exchange of, 73–4
   cycling of, 91
   deficiency of, 54
   effect of drainage, 102

Index

molybdenum (*contd.*)
  essential element, 2
  form of, 69, 73–4
montmorillonite, 19, 25
mudstone, 17
muscovite, 12, 15, 19, 23
mycorrhiza, 45

Nernst Equation, 84
nickel
  as a pollutant, 152
  as an essential element, 79
  in rocks, 11
nitrate
  as a pollutant, 110, 149
  production in soil, 77
  reduction of, 51, 84, 85
  uptake by foliage, 93, 160
nitrate reductase, 51
nitric acid
  as a pollutant, 155, 157–62
  nitrogen uptake from, 160
nitrification, 7, 122, 140, 160–1
nitrite, 77, 129
nitrogen
  cycling of, 3, 35, 39, 93–4, 109, 160–2
  effect of pH on, 77–8
  essential element, 2
  fertilisers, 107, 108
  fixation, 8, 93, 94
  global losses, 109
  mineralisation, 78, 122, 160–1
  organic, 35, 77–8
  status assessment, 109, 172
  unavailable pools of, 105
nitrous oxide, 7
  see also denitrification
nontronite, 19
norite, 12, 141
nuclear magnetic resonance, 39, 56, 172
nutrient uptake, 100

obligate anaerobes, 51
obsidian, 12, 13
olivine, 11, 12, 15, 19
organic farming, 46
organic matter
  accumulation of, 162
  adsorption of pesticides, 163–7
  characterisation of, 42–4
  components of, 28
  effect on structure, 44–7
  protected, 28, 39, 41–2, 48
  replacement of, 46–7
  turnover of, 47–9

organic pollutants, 153
orthoclase, 12, 15, 19, 21
outer-sphere complex, 59
overland flow, 131
oxygen, 50
  cycling of, 92
  reduction of, 84
  replenishment, 50
ozone, 92

particle size
  classification, 64
  distribution, 3, 62
peat, 16, 49, 142
  acidification of, 159–60
pectic acid, 33
pectin, 33
pectin esterase, 33
pegmatite, 13
peridotite, 11, 13
permease, 32
persistent binding, 46–7
pesticides
  adsorption by soils, 163–7
  effects on water quality, 150
pH
  changes in, 67–8, 119–21
  control of, 116
  definition of, 65
  duration curves, 134–5
  effects of pollution on, 157–60
  effects of trees on, 122, 147, 158
  effects on nutrient uptake, 118
  effects on nutrients, 67, 68–79
  effects on plants, 116–19, 120
  measurement of, 66–7, 114–16
Phanaerochaete, 35
phosphorus
  budgets, 4
  cycling of, 3, 35, 39, 91
  effect of pH on, 74–7
  essential element, 2, 8
  fertilisers, 107–8, 111–12
  fixation of, 74–5
  forms of, 69, 74–7
  in sludge, 4, 152–3
  labile pool, 125
  status assessment, 112, 172
  unavailable pools of, 105
photosynthesis, 1, 8
pitchstone, 13
plagioclase, 11, 12, 15
  porphyries, 12
plant breeding, 81, 106
plant residues, *see* litter
podzolisation, 142

pollution, 4, 5, 172
  heavy metal, 152–3
  in freshwaters, 128, 151
  of groundwater, 110
  of soil, 151–67
  organic, 153
polysaccharides, 41–2, 45
potassium
  cycling of, 68
  effect of pH on, 69
  essential element, 2, 8
  fertilisers, 107, 112–13
  fixation of, 132
  in interlayers, 23–4, 26
  outputs in rivers, 18, 142
  status assessment, 172
  unavailable pools of, 105
potato scab, 104
precipitation
  collection of, 97
  effects on fertiliser loss, 110
  effects on water quality, 131–4, 136–8
  interactions with soil, 4, 128–36, 143–7
  simulation of, 99, 159
  types of, 4
Pseudomonas, 31, 33, 51
pumice, 13
pyrite, 11
pyrolusite, 54
pyrophyllite, 23
pyroxene, 19, 20, 24
pyrrhotite, 11
pyruvate, 53

quartz, 12, 15, 16, 19
quartzite, 17, 158

radioisotopes, 30, 38, 82, 166
  as pollutants, 155–7
RAINS Model, 161
Ratio Law, 61
redox reactions, 51, 83–90
remote sensing, 172
respiration, root, 3, 30
rhizosphere, 29
  pH of, 125
rhyolite, 12, 13
rock
  nature of, 10–16
  weathering of, 8, 10, 17–18
rock salt, 16
root exudates, *see* exudates
roots, 29
rutile, 19

saline-sodic soils, 83

saline soils, 79–82
salinity
  effects on plants, 81
  in irrigated land, 100, 148
sandstone, 16, 158
saponite, 19
saprophytes, 36
sea salts, 4, 138
sediment core analysis, 155
sedimentary rocks, 14, 16, 17
selenium
  as an essential element, 78
  cycling of, 95
  forms of, 69
  toxicity, 2, 78
serpentine, 101
serpentinite, 11, 13, 25
sewage sludge, 4, 152–3
shale, 16, 17
shifting cultivation, 104
siderite, 16
silica, 14
  as cementing agent, 16
silicate structures
  chains, 19–20
  frameworks, 21–2
  sheets, 20–1
  tetrahedra, 18–19
silicon in rocks, 11
siltstone, 17
SMART Model, 146, 161
snow, 97, 134, 146, 147
sodic soils, 82–3
sodium
  cycling of, 68
  outputs in rivers, 18, 142
soil
  associations, 102
  mapping, 102, 143, 162
  series, 102
  solution, 7, 113, 121
  water tension, 130
soluble salts, 131
speciation, 171–2
springtails, 36
stability, *see* structure, weathering
stilbite, 19
stones, 12
strontium as a pollutant, 155–6
structure stability, 3, 44–7
succession, microbial, 36–8
sulphate
  adsorption of, 95, 96, 158–61
  reduction of, 51, 84, 85, 90
sulphides
  in rocks, 11

sulphides (*contd.*)
  precipitation of, 90
sulphur
  cycling of, 3, 35, 39, 93–4, 96
  effect of pH on, 78
  essential element, 2
  forms of, 69, 78
  organic, 78
sulphur dioxide, 4, 93, 95
sulphuric acid
  as a pollutant, 155, 157–62
  of volcanic origins, 4
surface area, 63–4
surface processes, 169
  see also anion exchange, cation
    exchange, pesticide adsorption
surface runoff, 131
sustainable ecosystems, 170
swelling lattices, 25, 26
syenite, 12
symbiosis, 8

talc, 23
temperate forest, 29
temperature
  effects of organic matter on, 57
  effects of pollution on, 153
  effects on water quality, 138–9
temporary binding, 45
tin, 79
TOPMODEL, 143
tourmaline, 54
toxicity, 152
transient binding, 45
trees, effect on pH, 122, 147, 158
Trichoderma, 31

trioctahedral sheet, 22, 26
tropical rainforest, 29

ultrabasic rocks, 11, 13, 23, 26

vanadium, 79
Van der Waals' forces, 163
vermiculite, 19, 26, 60
volcanic activity, 4

Walker Branch watershed, 157
water
  cycling of, 91–2, 130
  quality, 4, 122, 129–50
  retention in soil, 3, 8
  table, perched, 131, 134
weathering
  geochemical, 17–18, 132, 157, 169
  physical, 8
  rates of, 17, 23, 157, 162
  sequence of, 18, 19
white rot fungi, 32, 35

xenobiotics, 54

zinc
  as a pollutant, 152
  as substitute for aluminium, 23
  cycling of, 91
  effect of pH on, 70–2
  essential element, 2
  in rocks, 11
  speciation of, 55
zircon, 19
zymogenous decomposers, 36–7

Printed in the United Kingdom
by Lightning Source UK Ltd.
101686UKS00001B/185